Nonlinear Maximum Entropy Spectral Analysis Methods for Signal Recognition

ELECTRONIC & ELECTRICAL ENGINEERING RESEARCH STUDIES

PATTERN RECOGNITION & IMAGE PROCESSING RESEARCH STUDIES SERIES

Series Editor: **Dr. Josef Kittler,**
Rutherford Appleton Laboratory, Didcot, Oxon., England

1. Statistical Pattern Classification using Contextual Information
   K-S Fu *and* **T. S. Yu**

2. Kernel Discriminant Analysis
   D. J. Hand

3. Nonlinear Maximum Entropy Spectral Analysis Methods for Signal Recognition
   C. H. Chen

Nonlinear Maximum Entropy Spectral Analysis Methods for Signal Recognition

Professor C. H. Chen

Professor of Electrical Engineering,
Southeastern Massachusetts University,
North Dartmouth, Massachusetts, U. S. A.

RESEARCH STUDIES PRESS
A DIVISION OF JOHN WILEY & SONS LTD
Chichester · New York · Brisbane · Toronto · Singapore

RESEARCH STUDIES PRESS

Editorial Office:
58B Station Road, Letchworth, Herts. SG6 3BE, England

Library of Congress Cataloging in Publication Data:

Chen, C. H.
Nonlinear maximum entropy spectral analysis methods for signal recognition.
(Pattern recognition & image processing series; 3)
Includes bibliographical references and index.
1. Signal processing. 2. Spectrum analysis.
3. Entropy (Information theory) I. Title. II. Series.
TK5102.5.C474 1982 621.38'0433 82-8630
ISBN 0 471 10497 3 AACR2

British Library Cataloguing in Publication Data:

Chen, C. H.
Nonlinear maximum entropy spectral analysis methods for signal recognition.—(Pattern recognition and image processing research studies series)
1. Signal processing—Digital techniques
2. Spectral theory (Mathematics)
I. Title II. Series
621.38'0433 TK5102.5

ISBN 0 471 10497 3

Printed in Great Britain

Table of Contents

APPENDICES

Preface

In the past decade, there has been strong interest and much activity in developing high resolution power spectrum analysis techniques especially for short data records. The Burg's maximum entropy spectral analysis is a good example of such effort. While much work has been centered on the maximum entropy idea, there are many other elegant approaches to this important topic in spectral analysis. However theoretical performances are in most cases not supported by the computer results. This book is a research monograph based primarily on the extensions of the Burg's maximum entropy spectral analysis. The method due to P.F. Fougere that makes use of a nonlinear error minimization procedure is examined in great detail. The multichannel and two-dimensional maximum entropy spectral analysis methods are also carefully studied. These methods are nonlinear as they are data-adaptive. The spectral analysis is examined in the context of signal recognition. The highly accurate nonlinear spectral analysis methods presented provide effective features for automatic or interactive interpretation and classification of digital waveforms or images. The book places a major emphasis on computer implementation and application examples of the spectral analysis methods. The algorithms can all be efficiently implemented with the minicomputer. The procedures utilized are iterative and convergent to correct solutions. In spite of other new and super resolution spectral analysis techniques, it is my strong belief that the nonlinear methods, as supported by extensive computer results in the book, will have useful applications in many practical signal recognition and analysis systems.

I am indebted to Dr. Paul F. Fougere who introduced me to his algorithm, my graduate students Mr. Chihsung Yen and Mr. Gia-Kinh Young for programming assistance, and Dr. Josef Kittler for carefully reading the preliminary manuscript.

C. H. Chen

To WANDA,

Ivan, and Christopher

CHAPTER 1
Introduction

1. Spectral Analysis and Signal Recognition

Accurate spectral analysis of time series data is required in many physical sciences and engineering disciplines. The Fourier analysis via fast Fourier transform is now a basic tool in almost every spectral analysis technique. The Fourier analysis, however, is accurate only when the record length is long. When the data record is short as in many geophysical applications, for instance, conventional power spectrum estimation using smoothing and windowing procedures may provide poor resolution and undesired Gibb's phenomenon. The Burg's maximum entropy method of spectrum analysis considerably improves the spectral resolution for short records. The method makes no assumption of the data outside the time interval specified and is thus least committing to the unavailable data. Burg's method is data dependent and therefore, in this sense, is nonlinear. Generally speaking, the method enhances the peaky component of the spectrum. However, spectrum splitting and frequency shifting frequently occur for the sinusoidal signals especially.

Recently, Fougere [1] [2] proposed a new method which starting with the result of Burg's method, adopted a nonlinear error minimization procedure. The prediction error is as low as possible under a stability constraint. The name of the nonlinear procedure reflects the fact that the error power is a function of the sines of independent variables. This method along with a number of related nonlinear spectral analysis methods is examined in detail in this book. Furthermore, relationships among these methods are also established.

Many signal recognition systems or procedures use features extracted from the spectral domain. Thus accurate spectral information as can be provided by nonlinear methods is very important in the signal recognition context. Features of interest are parameters, such as the autoregressive model coefficients and reflection coefficients, as well as the signal power over certain frequency bands. Both feature extraction and classification performance of signal recognition systems using spectral analysis will be of main interest in this book.

2. Organization of the Book

After reviewing the Burg's univariate maximum entropy spectral analysis, Chapter 2 provides the mathematical details of the Fougere's nonlinear maximum entropy spectral analysis. Chapter 3 is concerned with the problem of minicomputer implementation of the Fougere's method, with illustrative examples particularly in signal recognition. Signal classification and resolution capabilities of the nonlinear spectral analysis methods and relationships among the methods are examined in Chapter 4. Chapter 5, and Chapters 6 and 7 deal respectively with the multichannel (multivariate) maximum entropy spectral analysis and the two-dimensional maximum entropy spectral analysis with signal recognition examples involving both time series and imagery data. Chapter 8 gives a detailed bibliography on the subject. Major computer programs are documented in the Appendices.

3. Bibliographical Notes

An extensive list of references on the maximum entropy spectral analysis is given in [3]. A more complete list of publications on this topic and related methods and on applications of these methods is compiled in Chapter 8. These lists clearly indicate a strong research interest in the nonlinear spectral analysis in general and the maximum entropy methods in particular. Several recent books on this general subject are listed as Refs. 4-8 at the end of this chapter.

REFERENCES

1. Fougere, P. F., "A solution to the problem of spontaneous line splitting in maximum entropy power spectrum analysis," Journal of Geophysics Research, vol. 82, pp. 1051-1054, 1977.
2. Fougere, P. F., "A solution to the problem of spontaneous line splitting in maximum entropy power spectrum analysis of complex signals," Proc. of the RADC Spectrum Estimation Workshop, May 1978.
3. Chen, C. H., "A review of geophysical signal analysis and recognition," Proc. of the Second International Symposium on Computer-aided Seismic Analysis and Discrimination, N. Dartmouth, MA, August 1981, pp. 144-152.
4. Chen, C. H., "Digital Waveform Processing and Recognition," CRC Press, Boca Raton, FL, Jan. 1982.
5. Childers, D. G., (editor), "Modern Spectrum Analysis," IEEE Press, New York, 1978.
6. Findley, D. F., (editor), "Applies Time Series Analysis," Academic Press, New York, 1978.
7. Haykin, S., (editor), "Nonlinear Methods of Spectral Analysis," Springer-Verlay, Berlin, Heidelberg, New York, 1979.
8. Robinson, E. A., and S. Treitel, "Geophysical Signal Analysis," Prentice-Hall, New York, 1980.

CHAPTER 2

The Univariate Maximum Entropy Spectral Analysis

1. Introduction

In this chapter, the basic principles of the univariate maximum entropy spectral analysis will be introduced with a particular reference to the Burg's method and the improvement of the Burg's method due to Fougere. Given a record $\{x_n\}$ of length N from a stationary time series of zero mean, the sample autocorrelation function $R_x(k)$ can be computed as

$$R_x(k) = \frac{1}{N} \sum_{i=1}^{N-|k|} x_i \, x_{i+k}^* \tag{1}$$

In general, the data considered is complex and "*" denotes the complex conjugate. The Fourier transform of $R_x(k)$ is an estimate of the spectral density $S_x(f)$. The entropy (or strictly speaking the entropy rate) of the process is

$$H = \frac{1}{4B} \int_{-B}^{B} \log S_x(f)df \tag{2}$$

where $S_x(f) = T \sum_{k=-\infty}^{\infty} R_x(k)\exp(-j2\pi kfT)$

Here T is the sampling period and it is assumed that the time series is essentially limited to the frequency band $-B \leq f \leq B$. Because of the finite number of samples, only a finite number, say 2K + 1 with $K < N$, of autocorrelation values are known. We can maximize Eq. (2) with respect to the unknown $R_x(k)$, $|k| \geq K + 1$, subject to the constraint that the estimated spectral density, denoted as $\hat{S}_x(f)$ must be consistent with the known autocorrelations $R_x(0), R_x(1), \ldots, R_x(K - 1)$, i.e.,

$$R_x(k) = \int_{-B}^{B} \hat{S}_x(f)\exp(j2\pi kfT)df, \quad 0 \leq k \leq K \quad .$$

Here the estimated $\hat{S}_x(f)$, called the maximum entropy spectral density estimate, expresses maximum uncertainty with respect to the unknown information but is consistent with the known information. Derivation of $\hat{S}_x(f)$ is available in [1-3] and the result is

$$\hat{S}_x(f) = \frac{P_m}{B\left|1 + \sum_{k\ 1}^{m} g_k \exp(-j2\pi fkT)\right|^2} \tag{3}$$

where P_m and g_k are respectively the output power and the coefficients of a prediction error filter which is examined in the next section. Equation (3) is the same as the spectrum of an autoregressive (AR) model of the signal. It is interesting to note that the representation of a stochastic process by an AR model is that representation that exhibits the maximum entropy.

2. Prediction-Error Filter

Suppose the time series $\{x_n\}$ of length N is applied as input to a linear digital filter of impulse response $\{h_n\}$ of length m. Let $\{y_n\}$ and $\{d_n\}$ represent the resulting output and the desired output respectively which both have length m + N - 1. The error between the filter output and the desired output is given by

$$e_n = d_n - y_n$$

The Wiener-Hopf equation that minimizes the mean square value of the error e_n is known in matrix form as

$$R_x h_o = R_{dx} \quad \text{or} \quad h_o = R_x^{-1} R_{dx} \tag{4}$$

where h_o is called the Wiener filter and

$$R_x = \begin{bmatrix} R_x(0) & R_x(-1)\ldots R_x(1-m) \\ R_x(1) & R_x(0)\ldots\ldots R_x(2-m) \\ \cdot & \\ \cdot & \\ R_x(m-1) & R_x(m-2)\ldots R_x(0) \end{bmatrix} \quad h_o = \begin{bmatrix} h_{01} \\ h_{02} \\ \cdot \\ \cdot \\ h_{0m} \end{bmatrix} \quad R_{dx} = \begin{bmatrix} R_{dx}(0) \\ R_{dx}(1) \\ \cdot \\ \cdot \\ R_{dx}(m-1) \end{bmatrix}$$

with $R_x(k) = E(x_n x^*_{n-k})$

$R_{dx}(k) = E(d_n x^*_{n-k})$

The minimum value of the error power is

$$R_{min} = R_d(0) - R_{dx}^H h_o \tag{5}$$

where the subscript H denotes the Hermitian transpose.

Consider the special case of a Wiener filter that predicts the sample of $\{x_n\}$ one time unit ahead, i.e.

$$d_n = x_{n+1}$$

Then the cross correlation between the input and the desired output becomes

$$R_{dx}(k) = E(x_{n+1+k}x_n^*) = R_x(k+1)$$

Equation (4) can then be written as

$$\sum_{i=1}^{m} h_{o,i}R_x(k-i) = R_x(k), \quad k = 1,2,\ldots,m \tag{6}$$

With the output of the predictive filter delayed by one time unit, the error e_n can be written as

$$e_n = d_n - y_n = x_n - \sum_{k=1}^{m} h_{o,k}x_{n-k} = \sum_{k=o}^{m} g_k x_{n-k} \tag{7}$$

Equation (7) defines a prediction error filter (PEF) of order m, with input $\{x_n\}$ and output e_n. The filter coefficient g_k is defined by

$$g_o = 1; \quad g_k = -h_{o,k}, \quad k=1,2,\ldots,m \tag{8}$$

where $h_{o,k}$ can be computed from Eq. (6). Note that the length of the impulse response of PEF is greater than that of the impulse response $h_{o,k}$ of the predictive filter by one. The output power of the PEF is the minimum error power which is termed the prediction error power P_m. For the one-step prediction considered above, $d_n = x_{n+1}$, and hence

$$R_d(0) = R_x(0)$$

$$R_{dx}^H = [R_x^*(1) \; R_x^*(2) \ldots R_x^*(m)]$$

$$= [R_x(-1) \; R_x(-2) .. R_x(-m)]$$

Thus the prediction error power is determined from Eq. (5) as

$$P_{min} = \sum_{k=o}^{m} g_k R_x(-k) = P_m \tag{9a}$$

Also Eq. (6) can be written as

$$\sum_{k=o}^{m} g_k R_x(i-k) = 0, \quad i = 1,\ldots,m \tag{9b}$$

Equation (9) can then be expressed in matrix form as

$$\begin{bmatrix} R_x(0) & R_x(-1)\ldots\ldots R_x(-m) \\ R_x(1) & R_x(0)\ldots\ldots R_x(1-m) \\ \vdots & \\ R_x(m-1) & R_x(m-2)\ldots\ldots R_x(-1) \\ R_x(m) & R_x(m-1)\ldots\ldots R_x(0) \end{bmatrix} \begin{bmatrix} 1 \\ g_1 \\ \vdots \\ g_m \end{bmatrix} = \begin{bmatrix} P_m \\ 0 \\ \vdots \\ 0 \end{bmatrix} \tag{10}$$

3. The Levinson Recursion

In order to calculate the maximum entropy spectral density estimate $\hat{S}_x(f)$, we have to calculate the output power P_m of the prediction error filter and the corresponding filter coefficients g_k. A recursive procedure to calculate P_m and g_k can be established as follows.

Define a forward and backward prediction error filters as the PEF with the input data in normal (forward) and reverse orders respectively. The forward and backward filter outputs are then given as

$$e_{f,n}^{(m)} = \sum_{k=o}^{m} g_k^{(m)} x_{n-k} \tag{11}$$

and

$$e_{b,n}^{(m)} = \sum_{k=o}^{m} g_{m-k}^{(m)*} x_{n+k-m} \tag{12}$$

where the superscript m is for filter order. For a forward PEF of order m - 1, Eq. (9) is given by

$$\sum_{k=o}^{m-1} g_k^{(m-1)} R_x(i-k) = \begin{cases} P_{m-1}, & i = 0 \\ 0, & i = 1,\ldots,m-1 \end{cases} \tag{13a}$$

or in matrix form

$$\begin{bmatrix} R_x(0) & R_x(-1)\ldots\ldots R_x(1-m) \\ R_x(1) & R_x(0)\ldots\ldots R_x(2-M) \\ \vdots & \\ R_x(m-1) & R_x(m-2)\ldots\ldots R_x(0) \end{bmatrix} \begin{bmatrix} 1 \\ g_1^{(m-1)} \\ \vdots \\ g_{m-1}^{(m-1)} \end{bmatrix} = \begin{bmatrix} P_{m-1} \\ 0 \\ \vdots \\ 0 \end{bmatrix} \tag{13b}$$

The backward PEF of order m - 1 can be obtained by taking the complex conjugate of both sides of Eq. (13a) and recognizing that the prediction error power P_{m-1} is a real quantity and that $R_x^*(k) = R_x(-k)$. Also substituting m - 1 - k for k, and m - 1 - i for i, we have

$$\sum_{k=o}^{m-1} g_{m-1-k}^{(m-1)*} R_x(i-k) = \begin{cases} 0, & i = 0,1,\ldots,m-2 \\ P_{m-1}, & i = m-1 \end{cases} \tag{14a}$$

or in matrix form

$$\begin{bmatrix} R_x(0) & R_x(-1)\cdots\cdots R_x(1-m) \\ R_x(1) & R_x(0)\cdots\cdots R_x(2-m) \\ \vdots & \\ R_x(m-1) & R_x(m-2)\cdots\cdots R_x(0) \end{bmatrix} \begin{bmatrix} g_{m-1}^{(m-1)*} \\ g_{m-2}^{(m-1)*} \\ \vdots \\ 1 \end{bmatrix} = \begin{bmatrix} 0 \\ 0 \\ \vdots \\ P_{m-1} \end{bmatrix} \tag{14b}$$

We can combine Eqs. (13b) and (14b) to expand the number of PEF equations by one as follows:

$$\begin{bmatrix} R_x(0) & R_x(-1)\cdots\cdots R_x(1-m) & R_x(-m) \\ R_x(1) & R_x(0)\cdots\cdots R_x(2-m) & R_x(1-m) \\ \vdots & & \\ R_x(m-1) & R_x(m-2)\cdots\cdots R_x(0) & R_x(-1) \\ R_x(m) & R_x(m-1)\cdots\cdots R_x(1) & R_x(0) \end{bmatrix}$$

$$\left[\begin{bmatrix} 1 \\ g_1^{(m-1)} \\ \vdots \\ g_{m-1}^{(m-1)} \\ 0 \end{bmatrix} + r_m \begin{bmatrix} 0 \\ g_{m-1}^{(m-1)*} \\ \vdots \\ g_1^{(m-1)*} \\ 1 \end{bmatrix} \right] = \left[\begin{bmatrix} P_{m-1} \\ 0 \\ \vdots \\ 0 \\ \Delta_m \end{bmatrix} + r_m \begin{bmatrix} \Delta_m^* \\ 0 \\ \vdots \\ 0 \\ P_{m-1} \end{bmatrix} \right] \tag{15}$$

Then comparing Eq. (15) with Eq. (10), we obtain

$$g_k^{(m)} = g_k^{(m-1)} + r_m g_{m-k}^{(m-1)*}, \quad k = 0,1,\ldots,m \tag{16}$$

$$P_m = P_{m-1} + r_m \Delta_m^* \tag{17}$$

and

$$0 = \Delta_m + r_m P_{m-1} \tag{18}$$

The recursive procedure based on Eq. (16) is called the Levinson's recursion. The parameter r_m is called the partial reflection coefficient as it plays a similar role as the reflection coefficient in a terminated transmission line that have both forward and reflected waves. In the recursive formula of Eq. (16), note that for all values of m,

$$g_k^{(m)} = \begin{cases} 1, & \text{for } k = 0 \\ r_m, & \text{for } k = m \\ 0, & \text{for } k > m \end{cases}$$

Thus the Levinson's recursion algorithm is given by

$$g_k^{(m)} = g_k^{(m-1)} + g_m^{(m)} \; g_{m-k}^{(m-1)*} \tag{19}$$

Eliminating Δ_m from Eqs. (17) and (18) we have the following recursive formula for the prediction error power

$$P_m = P_{m-1}(1 - |r_m|^2) \tag{20}$$

which indicates that P_m is a nondecreasing function of the filter order m.

4. The Burg's Algorithm

For the PEF to be minimum phase, Burg has shown that the reflection coefficients must all lie in the range $|g_m^{(m)}| < 1$. This is also the condition for the filter to be stable. The main problem with the forward and backward methods described in the last section is that the computed reflection coefficients are not always less than 1 in magnitude and thus the filter stability is not guaranteed. One solution to this problem suggested by Burg is to minimize the mean prediction error power taken as the average of the forward and backward error powers, i.e.,

$$P_m = \frac{1}{2}(P_{f,m} + P_{b,m}) \tag{21}$$

where

$$P_{f,m} = (N - m)^{-1} \sum_{n=1}^{N-m} e_{f,n}^{(m)} \; e_{f,n}^{(m)*}$$

$$P_{b,m} = (N - m)^{-1} \sum_{n=1}^{N-m} e_{b,n}^{(m)} \; e_{b,n}^{(m)*}$$

The Burg's algorithm starts with m = 0 for which

$$P_o = R_x(0) = \frac{1}{N} \sum_{n=1}^{N} x_n x_n^*$$

and then computes P_1 from Eq. (21). The value of r_1 for which P_1 is a minimum is then determined by solving $\partial P_1/\partial r_1 = 0$. Continuing this process for higher integer values of m, we can deduce a general expression for the reflection coefficient given by [4]

$$r_m = \frac{-2 \sum_{n=m+1}^{N} e_{f,n}^{(m-1)} \; e_{b,n-1}^{(m-1)*}}{\sum_{n=m+1}^{N} [\,|e_{f,n}^{(m-1)}|^2 + |e_{b,n-1}^{(m-1)}|^2]}, \quad m = 1,2,\ldots \tag{22}$$

where the forward and backward prediction-error terms are given by Eqs. (11) and (12). Equation (22) indicates that, except for the minus sign, the reflection coefficient at stage m is equal to the normalized value of the cross-correlation of the forward prediction error $e_{f,n}^{(m-1)}$ and the delayed backward prediction error $e_{b,n-1}^{(m-1)}$, with both errors evaluated at stage m - 1. r_m computed in Eq. (22) is always less than or equal to one in magnitude as

$$(e_{f,n}^{(m-1)} - e_{b,n-1}^{(m-1)})(e_{f,n}^{(m-1)*} - e_{b,n-1}^{(m-1)*}) \geq 0$$

Thus the use of Burg's method for the reflection coefficients always assures the stability of the resulting prediction-error filter. Once r_m is determined, the remaining PEF coefficients can be determined from the Levinson recursion given by Eq. (19), which involves previously determined prediction error filters.

5. The Fougere Algorithm

The Burg's procedure is sufficient to ensure that the highest order filter coefficient has a magnitude less than 1 and that the filter is stable, i.e., the z transform of the filter has all its poles outside the unit circle. This procedure is not, however, necessary. If the PEF's (with leading "1" suppressed) of all orders 1,2,...,m are gathered in one complex triangular matrix G_m, we may write

$$G_m = \begin{bmatrix} g_1^{(1)} & & & \\ g_1^{(2)} & g_2^{(2)} & & \\ \vdots & & & \\ g_1^{(m)} & g_2^{(m)} & \cdots & g_m^{(m)} \end{bmatrix}$$

It is necessary and sufficient that all diagonal elements have a magnitude less than 1 and that the remaining elements are determined by using the Levinson algorithm. If the entire matrix is redetermined each time a new prediction error filter is found and if the diagonal elements are treated as independent variables restricted to have magnitude less than 1, then the minimization problem has more degree of freedom, and a lower minimum can be found iteratively by beginning with some set of coefficients as a first approximation. The Burg coefficients will usually serve as a reasonable starting values for the iteration. This procedure was developed by Fougere [5][6].

Another procedure suggested by Ulrych and Clayton [7] is to remove all constraints and solve the normal equations for the prediction error filters. This procedure is much faster than the Fougere's procedure. However, Burg [3] has pointed out that this procedure may not produce a true prediction error filter. As the filter order is increased, it becomes more and more likely that one or more of the poles will lie inside the unit circle. When that happens, the filter is unstable and the computed spectrum is not correct. The Fougere's procedure guarantees that the solution is always a true prediction error filter having all poles lying outside the unit circle. This will be true even if the method fails to converge because by construction the diagonal elements of the G_m matrix all have magnitudes less than 1.

To enforce the condition that the magnitudes of the diagonal elements (viz. the partial reflection coefficients) be less than 1, we can set [6]

$$g_j^{(j)} = u \sin\theta_j \exp i\phi_j \tag{23}$$

for complex signals, where $0 \leq \theta_j \leq \pi$, θ_j and ϕ_j are both real and u is a positive constant slightly less than unity. Typical choice of u is $1 - 10^{-6}$. u is adjusted so that all poles of the PEF will lie outside the unit circle and none lie on it. For real signals, we can set

$$g_j^{(j)} = u \sin\theta_j \tag{24}$$

The error power thus depends on θ_j for real signals and on θ_j and ϕ_j for complex signals, resulting in nonlinear operations for error minimization. For complex signals the nonlinear optimization procedure starts with the expression for variations in θ_j, ϕ_j,

$$\theta_j = \theta_j^o + \Delta\theta_j \tag{25a}$$

$$\phi_j = \phi_j + \Delta\phi_j \tag{25b}$$

Now expand the prediction errors $e_{f,n}$ and $e_{b,n}$ in a Taylor series about θ_j^o and ϕ_j^o and retain only the first two orders,

$$e_{f,n} = e_{f,n}^o + \sum_{j=1}^{m} \left(\frac{\partial e_{f,n}}{\partial\theta_j} \Delta\theta_j + \frac{\partial e_{f,n}}{\partial\phi_j} \Delta\phi_j \right)$$

and a similar expression for $e_{b,n}$. Then set $\partial P_m/\partial\Delta\theta_\alpha = 0$, and $\partial P_m/\partial\Delta\phi_\alpha = 0$, where $\alpha = 1,2,\ldots,m$, to find the minimum error power. This results in 2m linear equations [6] to solve for the corrections $\Delta\theta_j$ and $\Delta\theta_j$ by standard matrix methods. The corrections are then substituted into Eq. (25) and the process is repeated until the corrections become sufficiently small.

The gradient of P_m with respect to the independent variables θ_k and ϕ_k is written as

$$\nabla_{\theta,\phi} P_m = \sum_{\alpha=1}^{m} \left(\frac{\partial P_m}{\partial\theta_\alpha} \bar{\theta}_\alpha + \frac{\partial P_m}{\partial\phi_\alpha} \bar{\phi}_\alpha \right) \tag{26}$$

where $\bar{\theta}_\alpha$ and $\bar{\phi}_\alpha$ are unit vectors, and [6]

$$\frac{\partial P_m}{\partial\theta_\alpha} = (N - m)^{-1} \sum_{n=1}^{N-m} \left[\mathrm{Re}\left(e_{f,n} \frac{\partial e^*_{f,n}}{\partial\theta_\alpha}\right) + \mathrm{Re}\left(e_{b,n} \frac{\partial e^*_{b,n}}{\partial\phi_\alpha}\right) \right] \tag{27}$$

The gradient computed from Eq. (26) and (27) along with the value of P_m can be used to determine the minimum P_m. The two methods of finding the minimum P_m, viz. the minimization of P_m with respect to correction terms, and the use of gradient, can be used alternately to insure the convergence of the PEF coefficients. A flow graph of the Fougere algorithm is given in Appendix A.

It is simpler to consider signals which are real numbers. The procedure, as in the complex signal case, follows the variational method of Cain et.al. [7], with θ_j given by Eq. (25a). The prediction error is approximated by the first two terms of the Taylor series as

$$e_{f,n} = e^{o}_{f,n} + \sum_{j=1}^{m} \frac{\partial e_{f,n}}{\partial \theta_j} \Delta\theta_j$$

with a similar expression for $e_{b,n}$, and these are then substituted into Eq. (21) for the prediction error power,

$$P_m = 0.5(n - m)^{-1} \sum_{n=1}^{N-m} [(e^{o}_{f,n} + \sum_{j=1}^{m} \frac{\partial e_{f,n}}{\partial \theta_j} \Delta\theta_j)^2 + (e^{o}_{b,n} + \sum_{j=1}^{m} \frac{\partial e_{b,n}}{\partial \theta_j} \Delta\theta_j)^2] \tag{28}$$

Set $(\partial P_m / \partial \Delta_\alpha) = 0$ to find the minimum error power and then rearrange the resulting equation slightly to get

$$\sum_{j=1}^{m} \sum_{n=1}^{N-m} [\frac{\partial e_{f,n}}{\partial \theta_\alpha} \frac{\partial e_{f,n}}{\partial \theta_j} \Delta\theta_j + \frac{\partial e_{b,n}}{\partial \theta_\alpha} \frac{\partial e_{b,n}}{\partial \theta_j} \Delta\theta_j] = - \sum_{n=1}^{N-m} (e^{o}_{f,n} \frac{\partial e_{f,n}}{\partial \theta_\alpha} + e^{o}_{b,n} \frac{\partial e_{b,n}}{\partial \theta_\alpha}) \tag{29}$$

$$\alpha = 1,2,\ldots,m$$

The derivatives of $e_{f,n}$ and $e_{b,n}$ are obtained by differentiating Eqs. (11) and (12) as follows:

$$\frac{\partial e_{f,n}}{\partial \theta_j} = \sum_{k=1}^{m} x_{n-k} \frac{\partial g^{(m)}_k}{\partial \theta_j}$$

$$\frac{\partial e_{b,n}}{\partial \theta_j} = \sum_{k=1}^{m} x_{n+k-m} \frac{\partial g^{(m)}_{m-k}}{\partial \theta_j}$$

where the summation begins at k - 1 because $g_o = 1$. Now using Eq.(24) we have

$$\frac{\partial}{\partial \theta_j} = \frac{\partial g^{(j)}_j}{\partial \theta_j} \frac{\partial}{\partial g^{(j)}_j} = u\cos\theta_j \frac{\partial}{\partial g^{(j)}_j}$$

Then using the Levinson algorithm, we can find all necessary derivatives, i.e.

$$\frac{\partial g^{(i)}_k}{\partial g^{(j)}_j} = \frac{\partial g^{(i-1)}_k}{\partial g^{(j)}_j} + g^{(i)}_i \frac{\partial g^{(i-1)}_{i-k}}{\partial g^{(j)}_i} + g^{(i-1)}_{i-k} \delta_{ij}$$

For example, $\partial g_1^{(2)}/\partial g_1^{(1)} = 1 + g_2^{(2)}$, $\partial g_1^{(2)}/\partial g_2^{(2)} = g_1^{(1)}$, $\partial g_2^{(2)}/\partial g_1^{(1)} = 0$ and $\partial g_2^{(2)}/\partial g_2^{(2)} = 1$. Note also that $\partial g_k^{(i)}/\partial g_j^{(j)} = 0$ if $j > i$. Thus all partial derivatives can be computed and $\partial\theta_\alpha$ can be solved from Eq. (27) for $\alpha = 1,2,...m$. The minimum P_m is determined from Eq. (28).

If we need the gradient of P_m with respect to the independent variables θ_k, we can write

$$\nabla_\theta P_m = \sum_{\alpha=1}^{m} \frac{\partial P_m}{\partial \theta_\alpha} \bar{\theta}_\alpha \tag{30}$$

where $\bar{\theta}_\alpha$ is a unit vector and

$$\frac{\partial P_m}{\partial \theta_\alpha} = - (N - m)^{-1} \times (\text{right side of Eq. (27)})$$

In practice the gradient method is simple but unreliable. As discussed in the next Chapter, the Fletcher and Powell rapid descent method [8] and the conjugate gradient methods [9] perform much better even though more storage space is required.

6. PEF Order Selection

The selection of the order of the prediction error filter is based mainly on the data rather than on the particular algorithm minimizing the prediction error power. To choose a suitable filter order, we may use the so-called final prediction error (FPE) which is defined as the mean-square value of the prediction error [10]. For a prediction error filter of order $m + 1$, the final prediction error is

$$(\text{FPE})_m = \frac{N + (m + 1)}{N - (m + 1)} P_m \tag{31}$$

which is called the Akaike criterion. The idea is to fit the data with an autoregressive model. This criterion takes into account only the forward prediction error while the maximum entropy method considers both the forward and backward prediction errors. A suggested modification of Eq. (31) is

$$(\text{FPE})_m = \frac{N + 2\ (m + 1)}{N - 2\ (m + 1)} P_m \tag{32}$$

Empirical results tend to suggest, however, that the actual final prediction error lies in between the lower bound of Eq. (31) and the upper bound of Eq. (32).

REFERENCES

1. Haykin, S., editor, "Nonlinear Methods of Spectral Analysis", Springer-Verlag, Berlin, Heidelberg, New York, 1979, Chapter 2.
2. Ulrych, T. J., and T. N. Bishop, "Maximum entropy spectral analysis and autoregressive decomposition", Review of Geophysics and Space Physics, Vol. 13, pp. 183-200, February 1975.
3. Burg, J. P., "Maximum entropy spectral analysis", Ph.D. dissertation, Stanford University, Stanford, CA, 1975.
4. Burg, J. P., "A new analysis technique for time series data", presented at the NATO Advanced Study Institute on Signal Processing with Emphasis on Underwater Acoustics, Enschede, Netherlands, August, 1968.
5. Fougere, P. F., "A solution to the problem of spontaneous line splitting in maximum entropy power spectrum analysis", Journal of Geophysical Research, Vol. 82, pp. 1051-1054, 1977.
6. Fougere, P. F., "A solution to the problem of spontaneous line splitting in maximum entropy power spectrum analysis of complex signals", Proc. of the RADC Spectrum Estimation Workshop, May 1978.
7. Cain, J. C., S. J. Hendricks, R. A. Langel, and W. V. Hudson, "A proposed model for the international geomagnetic reference field - 1965", J. of Geomagnetics and Geoelectricity, Vol. 19, pp. 335-355, 1967.
8. Fletcher, R., and M. J. D. Powell, "A rapid descent method for minimization", Computer Journal, Vol. 5, No. 2, pp. 163-168, 1963.
9. Fletcher, R., and C. M. Reeves, "Function minimization by conjugate gradients", Computer Journal, Vol. 7, No. 2, pp. 149-154, 1964.
10. Akaike, H., "Fitting autoregressive models for prediction", Ann. Institute of Statistical Mathematics, Vol. 21, pp. 243-247, 1969.

CHAPTER 3
Computer Implementation

1. Implementation Using a Mini-Computer

The nonlinear maximum entropy spectral analysis method described in the previous chapter has been extensively examined by Fougere [1] and by this author [2] using a CDC 6600 computer. This computer provides the much needed computing power required for the algorithm. Double precision is frequently used in the spectral computation of sinusoids for which very sharp spectral peaks must be determined. If a minicomputer is used, the computation problem is more serious. For example, the word length of the PDP 11/45 minicomputer is only one-half of that of the CDC computer, while the memory capacity is typically limited to 28K bytes. The problem of small computer implementation of high resolution spectral analysis algorithms is very important as the improved microcomputer and microprocessor technology makes it possible to incorporate many sophisticated signal processing capabilities in very small machines. Also for real-time or near real-time applications, large computers may not be available.

After intensive research efforts, we have successfully implemented the nonlinear method on a minicomputer by utilizing double precision and making some modifications to the optimization procedures [3]. The resulting accuracy based on the sinewaves as test signals is nearly as good as that available from the CDC computer. A complete listing of the computer programs is given in Appendices B and C. Table 1 is a comparison of the software realization of the nonlinear method for the complex numbers at the input. For the real number input, 250 data points can easily be processed. However, the advantage of the maximum-entropy method over the fast Fourier-transform method diminishes as the number of data points becomes large. The following are some useful hints for the minicomputer implementation:

(1) A good choice of u (Equations 23 and 24, Chapter 2) is $u=1 - 10^{-5}$ for the minicomputer. For CDC 6600, a proper choice is $u=1 - 10^{-6}$.

(2) While single precision may be acceptable for Burg's method, double precision is necessary for the nonlinear method, implemented in the minicomputer. The use of the single precision

	OUR REALIZATION	FOUGERE'S REALIZATION
Computer	PDP 11/45	CDC 6600
Maximum array dimension of Fortran Compiler	16K blocks	virtually unlimited
Format of variables	Double-precision	Single-precision
Limit of number of input points	25 complex numbers	unlimited
Number of output points (spectrum)	251 points	601 points
Optimization procedure	Fletcher and Powell method	Both Fletcher and Powell method and Conjugate Gradient method
Filter coefficient accuracy	10 digits or more This accuracy is adequate for all practical purposes.	20 digits or more

TABLE I A Comparison of Software Realization of the Nonlinear Maximum Entropy Spectral Analysis with the minicomputer frequently results in an unstable or unreasonable prediction error filters for the nonlinear method.

(3) The number of iterations required for the final prediction-error reduction is usually limited to 30 and the round-off error accumulation is negligible. In fact, very little is gained in performance by increasing significantly the number of iterations.

For the hardware implementation, it is important to note that the microprocessor/microcomputer implementation of the Burg's method is now possible. The functional blocks, OPTM and FGR, referred to in Appendix A can be achieved with a microcomputer. Computation of SPEC can be performed with an Array Processor or a microcomputer. A 16-bit microcomputer is required to avoid the round-off error propagation.

The computer results presented in the following sections are based on the described implementation of Burg's and the nonlinear methods. Especially in the next section, the computer results clearly demonstrate that the maximum entropy spectral analysis greatly enhances the recognition of single and multiple sinewaves in the presence of noises.

2. Computer Results on Sinewaves Plus Noise

For notational convenience, let (x,y) denote x-point maximum entropy power spectrum for y filter weights (i.e. filter order y). Sampling rate is 20 samples per second for all sinewaves. Every division on the frequency scale is 0.5 Hz. Figure 1 provides a comparison of Burg and Fougere (nonlinear method) spectra for 1 Hz sinewave with six filter weights. Figure 2 is another comparison with eight filter weights. The improvement of the nonlinear method over the Burg's method is quite evident. Spectrum splitting is corrected by the nonlinear method. A good choice of the number of iterations is 10. Filter order of 8 is much better. For two sinewaves with different combination of initial phases, the Burg's method clearly demonstrates the spectrum shifting phenomenon which is corrected by the nonlinear method with enhanced resolution as demonstrated by Figs. 3 and 4. Here n(t) is the computer generated Gaussian random number of mean 0 and unit variance.

It would be of interest to examine the effect of additive noise. Figures 5, 6 and 7 are based on Burg's method for the 64-point sinewave plus noise at three different noise levels. As the noise level increases, spectrum splitting disappears but the spectrum shifting becomes evident. A filter order of 11 appears to be a good choice in Fig. 7. It should be noted that the scales of the vertical axis are different in all figures. The Tektronix display automatically normalizes the vertical scale by the peak value.

For complex sinewaves, the spectrum splitting phenomenon is much less evident. Fig. 8 is the Burg's power spectrum for the complex sinewave of 1 Hz, which does not have spectrum splitting. The Fougere method provides very little improvement in this case.

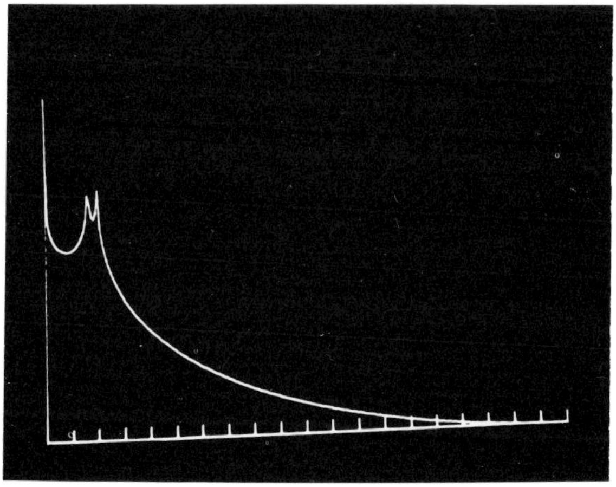

a) Burg's spectrum (16,6) in logarithmic scale

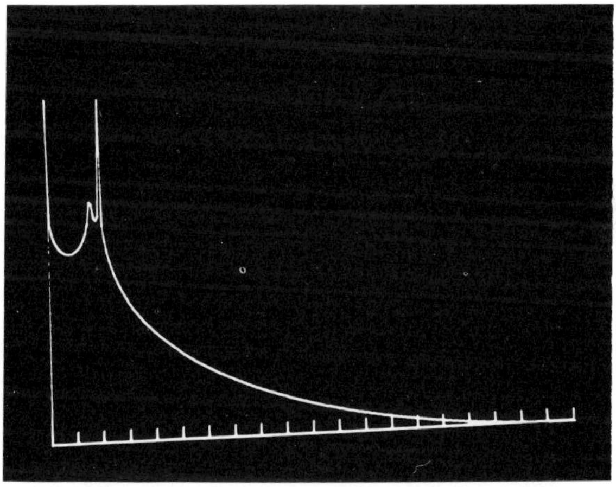

b) Fougere's spectrum (16,6) in logarithmic scale with 5 iterations

Figure 1 A comparison of Burg and Fougere spectra for 1 Hz sinewave with initial phase of 45°.

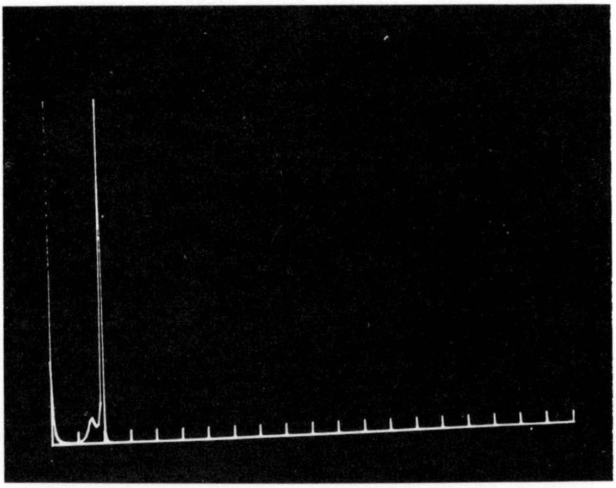

a) Burg's spectrum (16,8) in linear scale

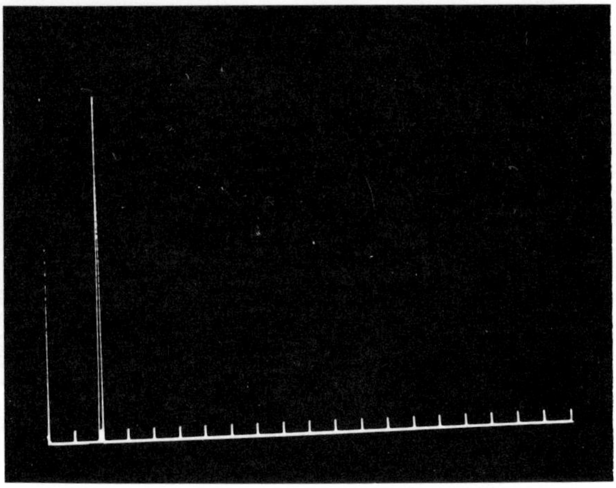

b) Fougere's spectrum (16,8) in linear scale with 10 iterations

Figure 2

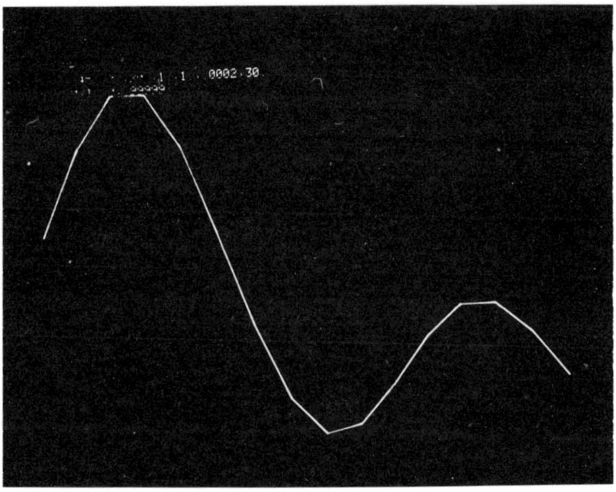

a) Plot of $\sin(2\pi t+45^{o})+\sin(4\pi t+0^{o})+0.0002n(t)$ for 16 points

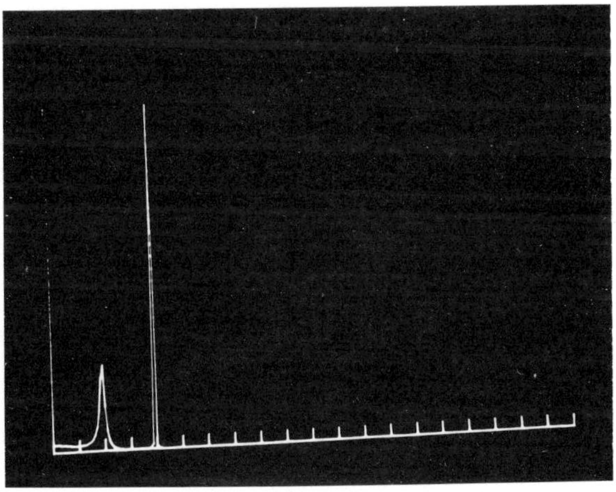

b) Burg's spectrum (16,8) in linear scale

Figure 3 A comparison of Burg and Fougere spectra for two sinewaves and additive noise.

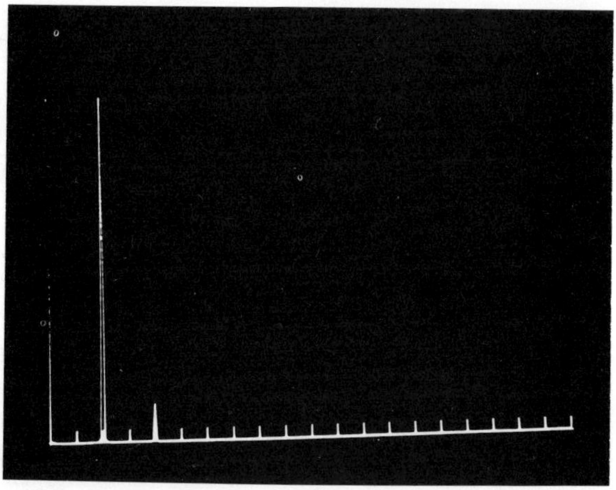

c) Fougere's spectrum (16,8) in linear scale with 20 iterations

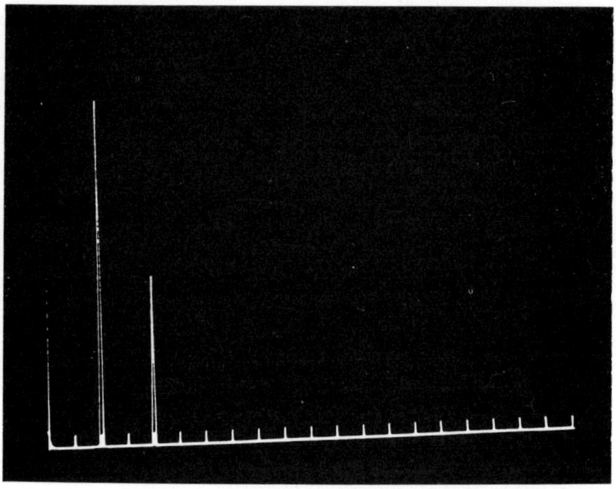

d) Fougere's spectrum (16,8) in linear scale with 40 iterations

Fig. 3 (continued)

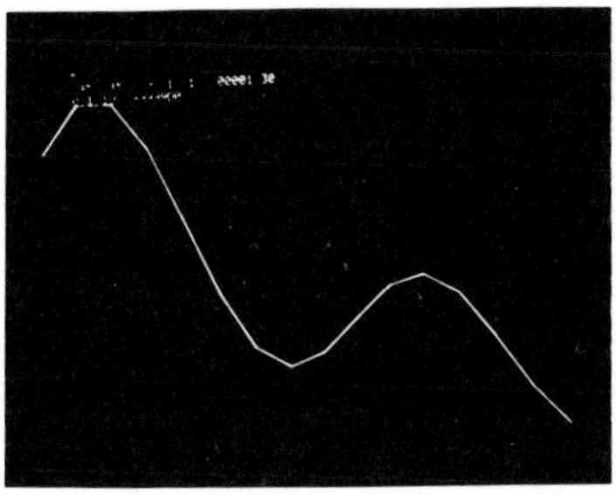

a) A plot of the waveform for 16 points

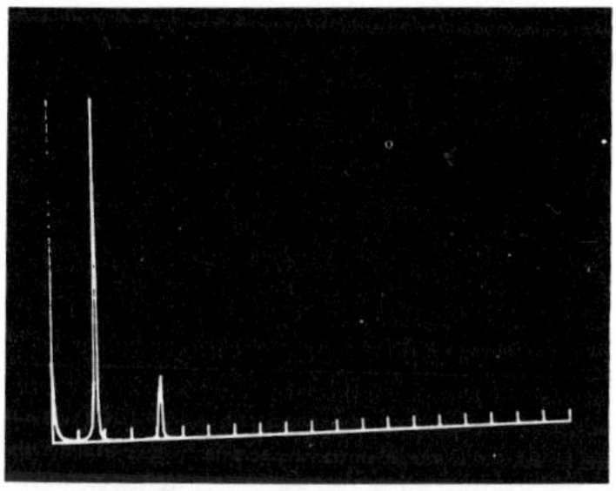

b) Burg's spectrum (16,8) in linear scale

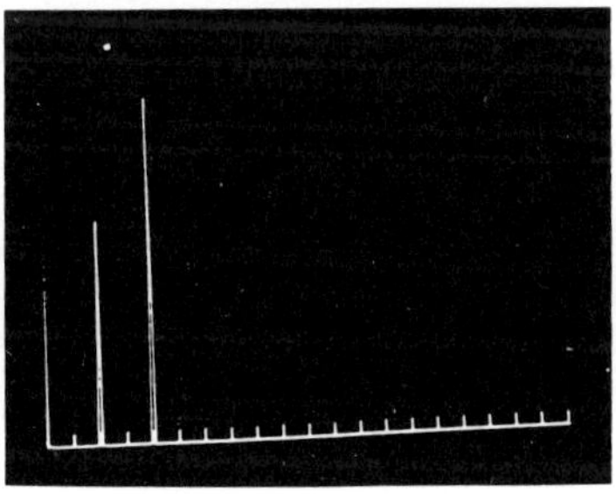

c) Fougere's spectrum (16,8) in linear scale for 10 iterations

Figure 4 A comparison of Burg and Fougere spectra for the waveform: $\sin(2\pi t+45^{\circ})+\sin(4\pi t+45^{\circ})+0.0001n(t)$.

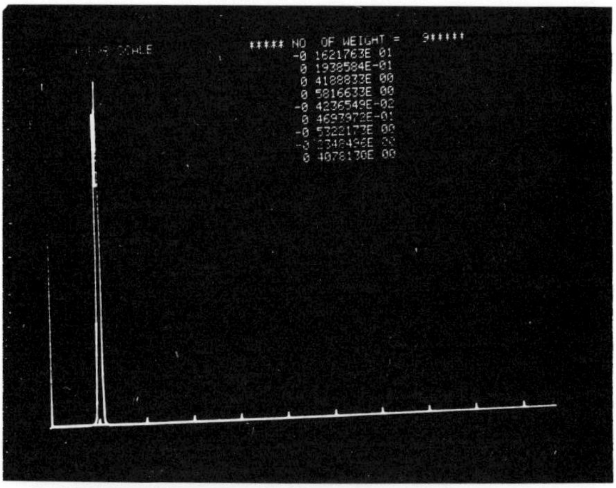

a) Spectrum in linear scale

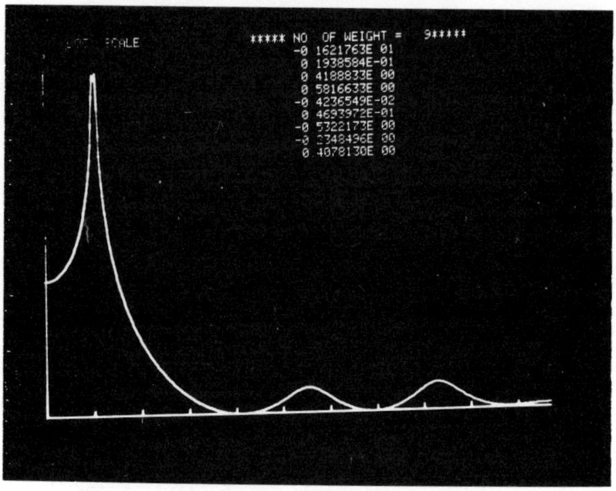

b) Spectrum in logarithmic scale

Figure 5 Burg's spectrum (64,10) for the waveform: $\sin(2\pi t+45^{o}) + 0.001n(t)$.

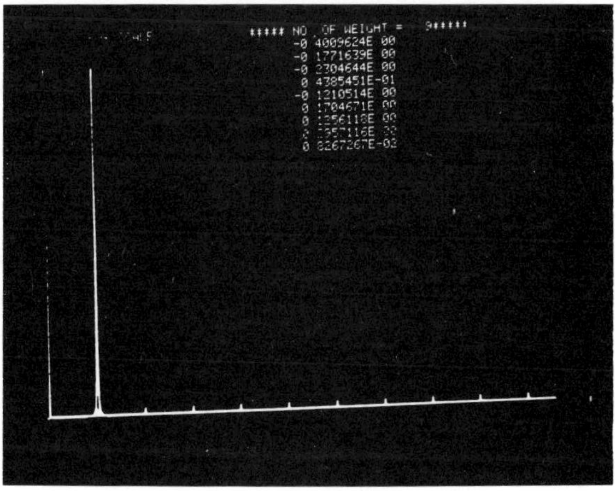

a) Spectrum in linear scale

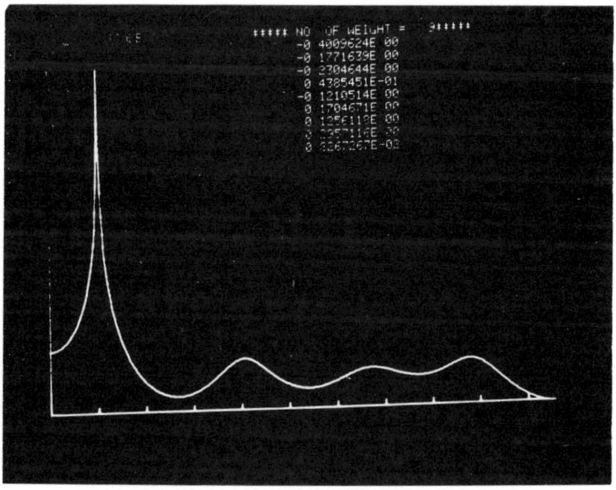

b) Spectrum in logarithmic scale

Figure 6 Burg's spectrum (64,10) for the waveform: $\sin(2\pi t+45^{o}) + 0.1n(t)$.

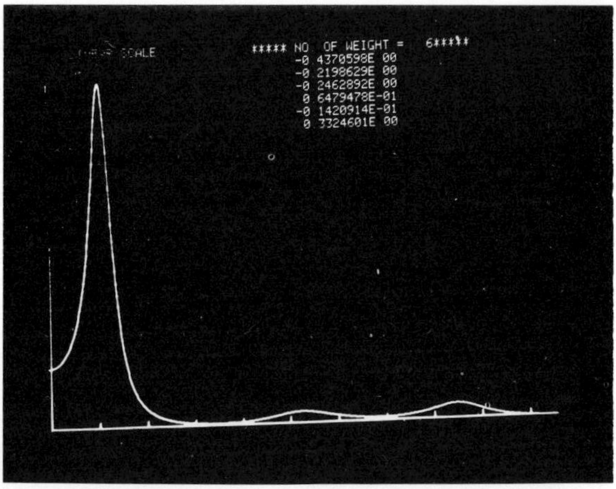

a) Spectrum in linear scale(64,7)

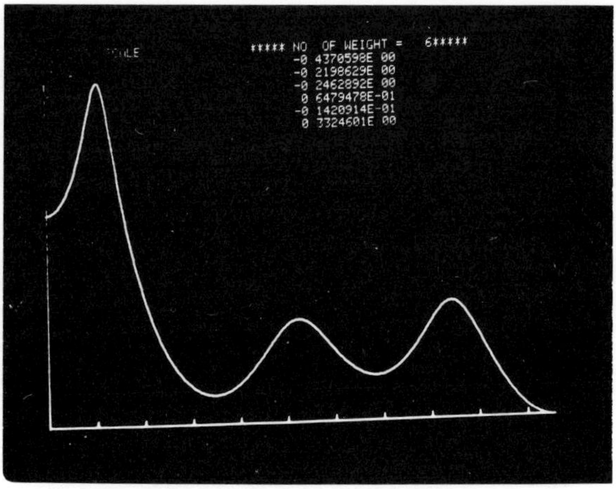

b) Spectrum in logarithmic scale(64,7)

Figure 7 Burg's spectra for 64-point waveform $\sin(2\pi t+45^{o}) + 0.5n(t)$ in three different filter orders.

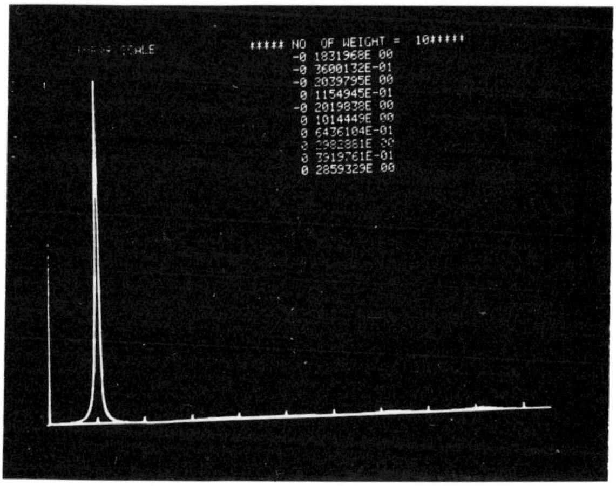

c) Spectrum in linear scale(64,11)

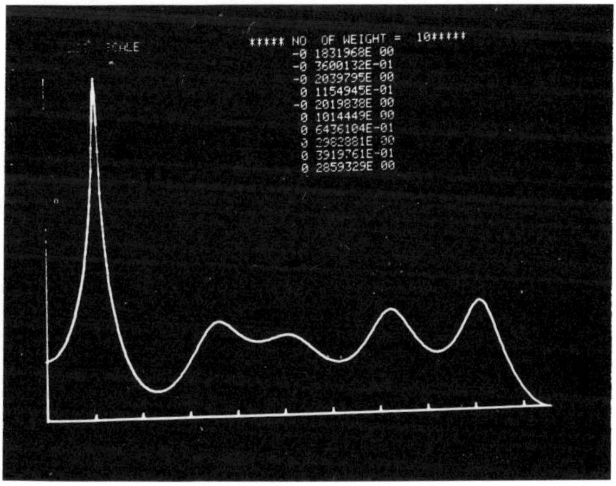

d) Spectrum in logarithmic scale(64,11)

Fig. 7 (continued)

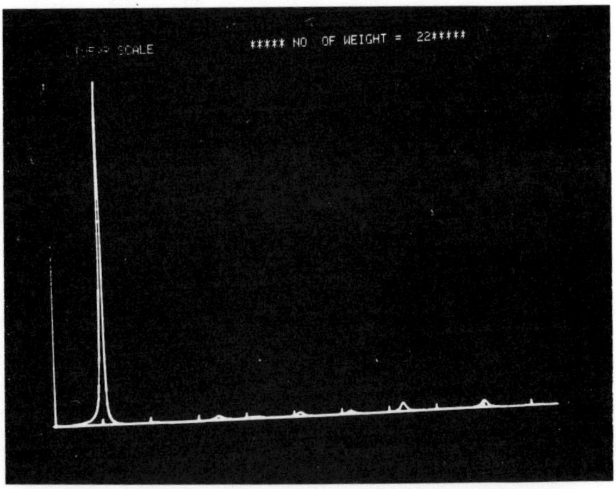

e) Spectrum in linear scale(64,23)

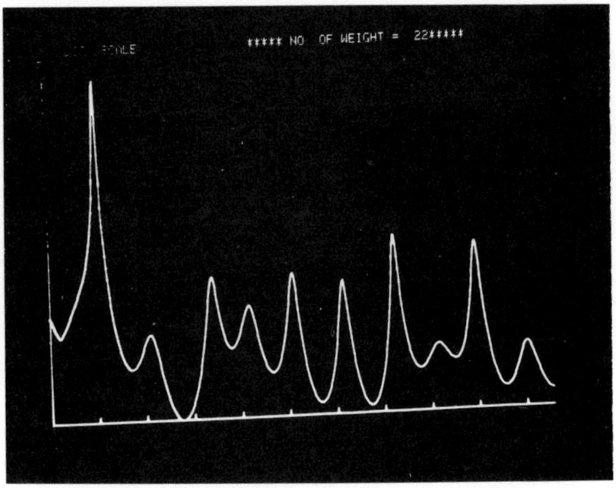

f) Spectrum in logarithmic scale(64,23)

Fig. 7 (continued)

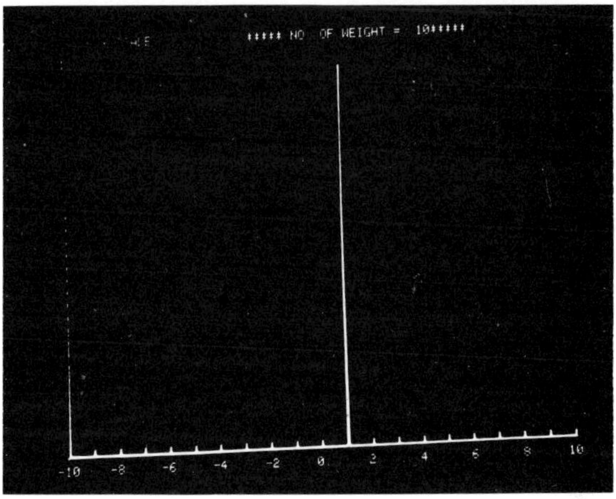

a) Spectrum in linear scale(32,10)

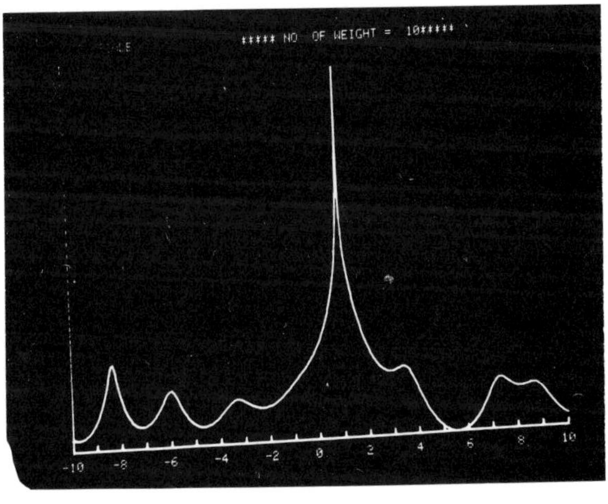

b) Spectrum in logarithmic scale(32,10)

Figure 8 Complex Burg's spectrum for 1 Hz sinewave.

3. Computer Results on Sunspots Data

Figure 9a is a plot of Zürich sunspot data from 1780 to 1979 with 1 point per year (sampling rate=1). The single precision Burg's program is used to obtain an accurate power spectrum of sunspot numbers for the 200 data points. The major periodicity (around 9 to 12) can be predicted accurately from the Burg's spectrum of Fig. 9b. Fig. 9c, also in linear scale, is the result of the nonlinear method based on 8 filter weights and only 20 data points (1960 to 1979). The improvement obtained with the nonlinear method is quite evident.

4. Computer Results on Radar Data [4]

Figures 10a and 10b are the two-channel radar data supplied by Rome Air Development Center for the 1979 RADC Spectrum Estimation Workshop. A 32-point complex number sequence is formed by these two channels. The spectrum of the nonlinear method is shown in Fig. 11a (linear plot) and Fig. 11b (logarithmic plot) for 10 filter weights. Burg's result is shown in Fig. 12a (linear plot) and Fig. 12b (logarithmic plot), also for 10 filter weights. A more detailed tabulation of the major frequency components outside the clutter bandwidth ($-0.167f_s$, $+0.167f_s$), where f_s is the sampling rate, is given in Fig. 13.

Frequency	Amplitude	Relative level
$+0.1836\ f_s$	0.2184	0 dB
$-0.3945\ f_s$	0.1203	-5.18 dB
$+0.2930\ f_s$	0.0873	-7.96 dB

True Answer

Frequency	Amplitude	Relative level
$+0.18359\ f_s$	5.48007	0 dB
$-0.39453\ f_s$	3.02523	-5.16 dB
$+0.29296\ f_s$	2.19424	-7.95 dB

Result of Nonlinear Method

Frequency	Amplitude	Relative level
$+0.18164\ f_s$	3.58956	0 dB
$-0.39453\ f_s$	1.45378	-7.85 dB
$+0.29492\ f_s$	1.40613	-8.14 dB

Result of Burg's Method

Figure 13

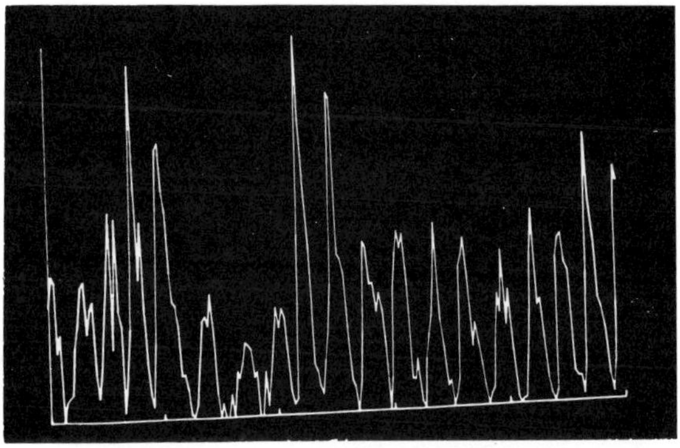

a) A plot of the data from 1780 to 1979

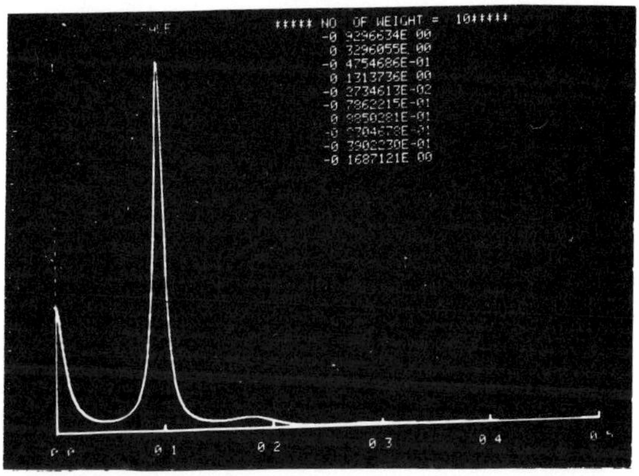

b) Burg's spectrum (200,11) in linear scale

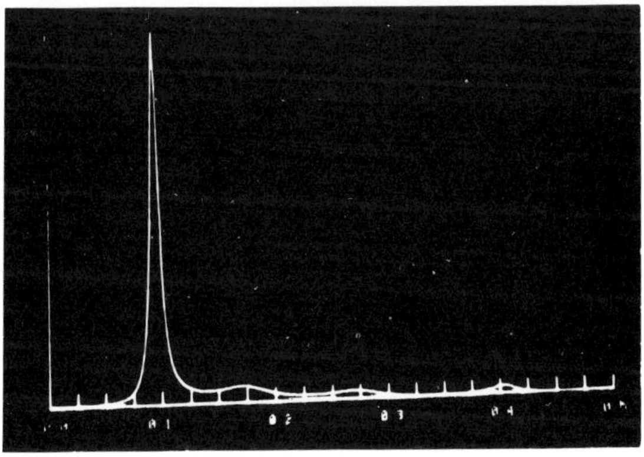

c) Fougere's spectrum(20,8) in linear scale for data from 1960 to 1979

Figure 9 Spectra of Zürich sunspot numbers.

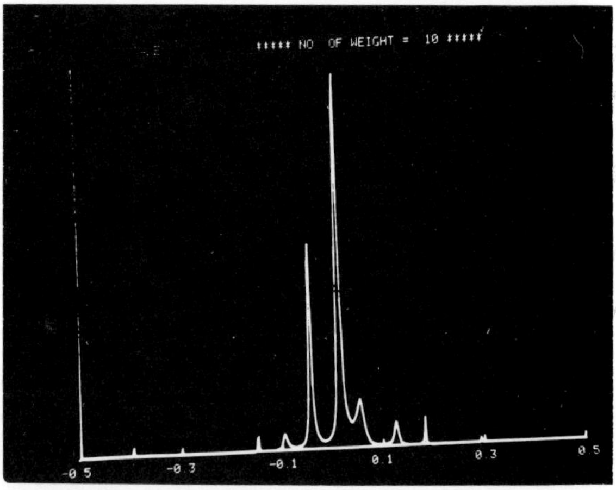

a) Spectrum in linear scale(32,10)

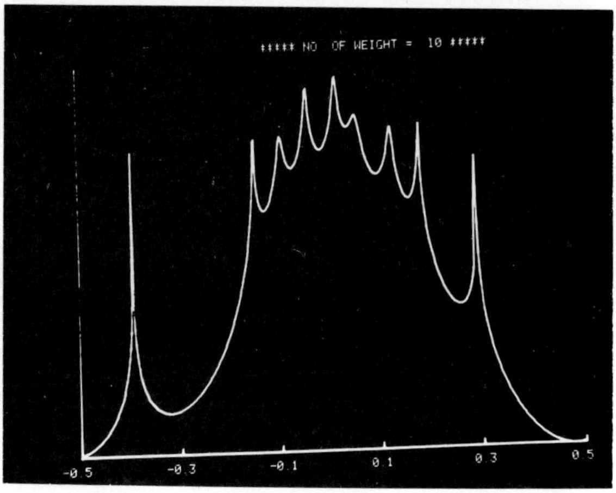

b) Spectrum in logarithmic scale(32,10)

Figure 11 Fougere's spectrum for the radar data.

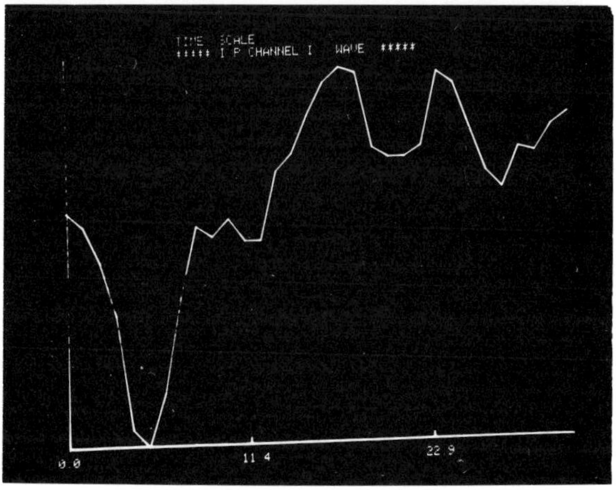

a) The I-channel data for real part

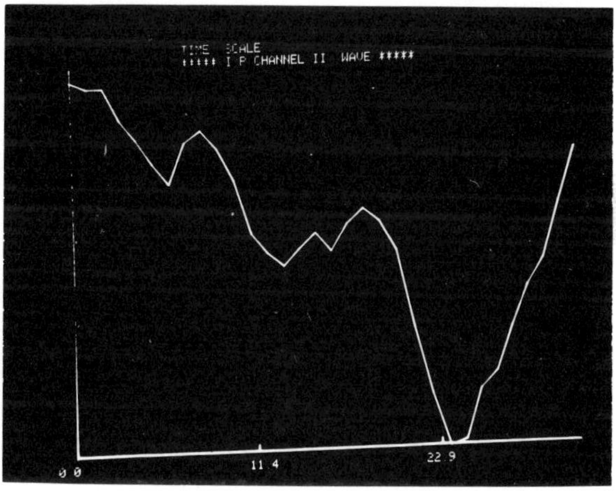

b) The Q-channel data for imaginary part

Figure 10 The 32-point complex radar data considered.

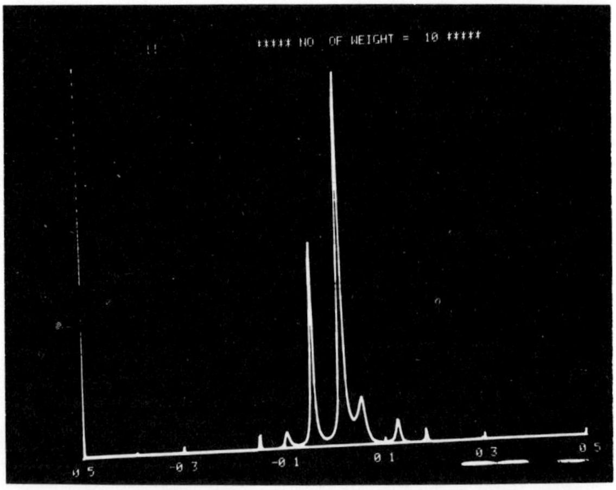

a) Spectrum in linear scale(32,10)

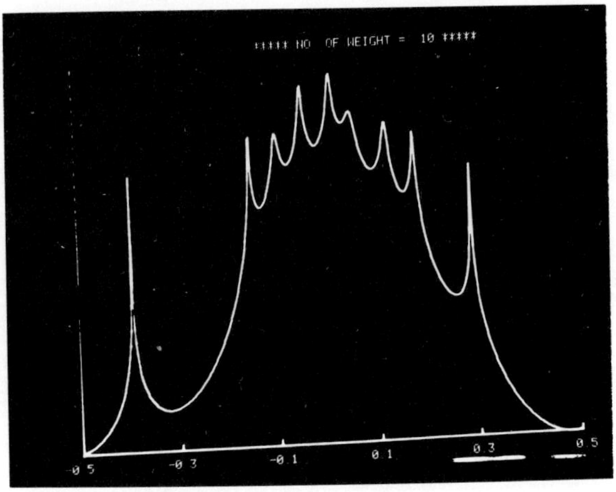

b) Spectrum in logarithmic scale(32,10)

Figure 12 Burg's spectrum for the radar data.

The nonlinear method which provides a result very close to the answer clearly is much better than the Burg's method. Here 30 iterations were used in the nonlinear method which appears to be an optimum number. The optimum filter order is around 10 or 11 as determined by Fig. 14 which shows the final prediction error (FPE) according to the Akaike criterion.

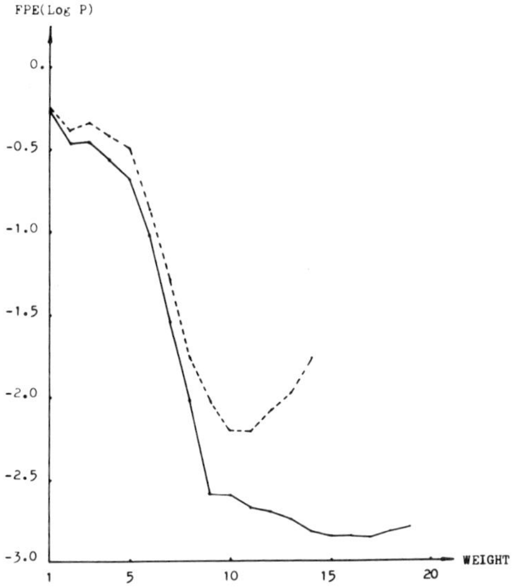

Figure 14 The final prediction error.

The solid curve is the lower bound and the dashed curve is the upper bound given by Eq. 32 of Chapter 2. It appears that the upper bound is better than the lower bound, and the true answer lies in between the two bounds. It is further noted that the maximum entropy spectral analysis makes it possible to recognize the target signals from the cluttered background, and to classify various targets.

5. Spectral Display by Computer Graphics

Computer graphics provides an effective three-dimensional display of the power spectral density for various parameters. Using the software we have developed, computer-graphics display for both Burg's and the nonlinear methods has been generated. Figure 15a is a plot of the January Zürich sunspot numbers for 200 data points for the period 1779 to 1978 (same as Fig. 9a). Figure 15b is the display of the corresponding Burg's spectrum for filter orders 10 to 58 and frequency interval 0.05 to 0.15 cycles per year. All spectrum plots are in logarithmic scale with base 10. The three numbers on the top of the photo are A=minimum spectral magnitude, B=maximum spectral magnitude, and C=difference between B and A, in log scale. For Fig. 15b, A=2.798, B=5.113, C-2.315. In all graphical displays of this section, 450

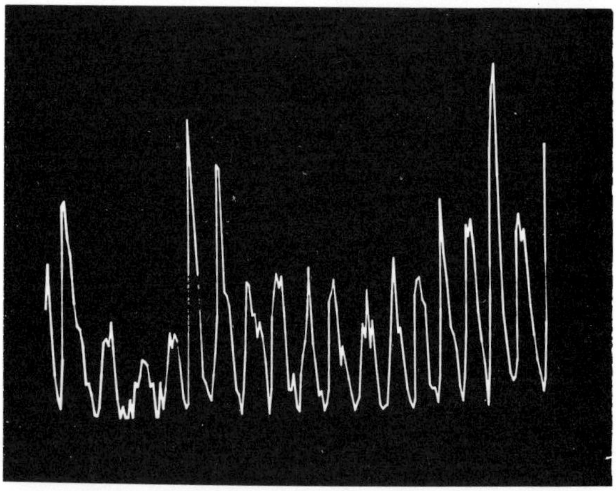

a) The January Zürich sunspot numbers, 1779 to 1978

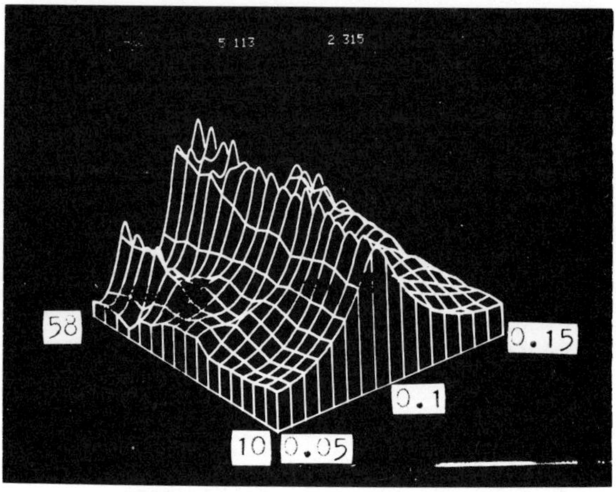

b) Corresponding Burg's spectra in 3-D plot

Figure 15

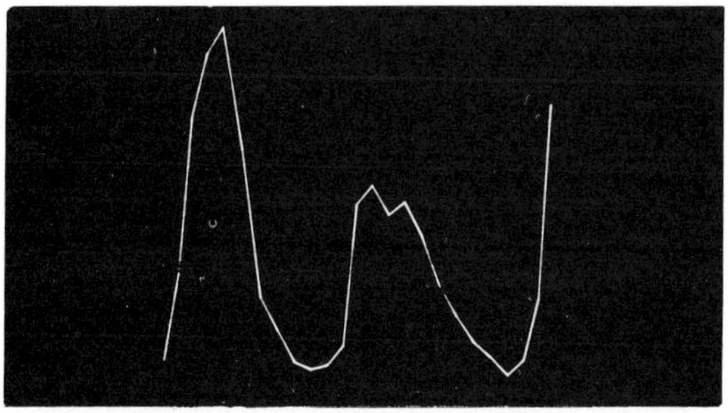

a) A 25-point section of Fig. 15a

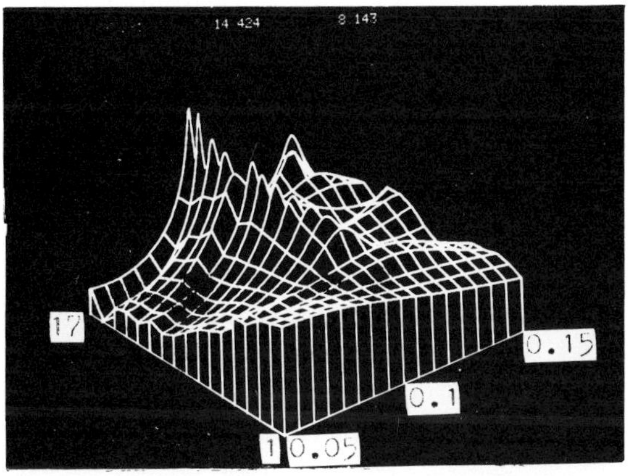

b) Burg's spectra in 3-D plot

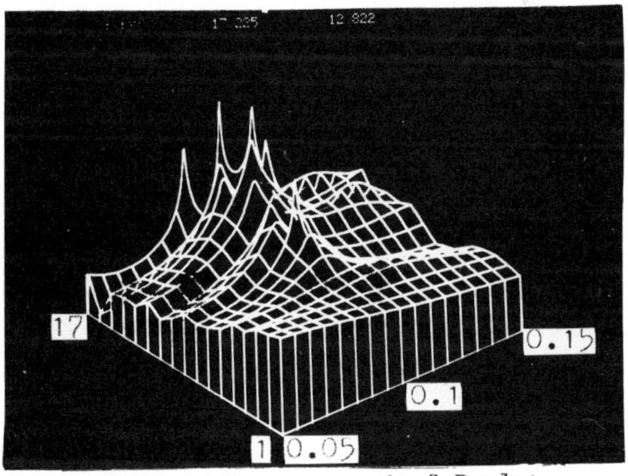

c) Fougere's spectra in 3-D plot

Figure 16

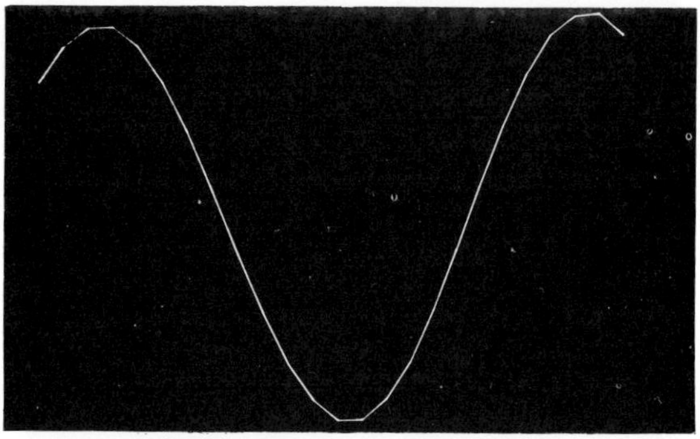

a) The sinewave considered

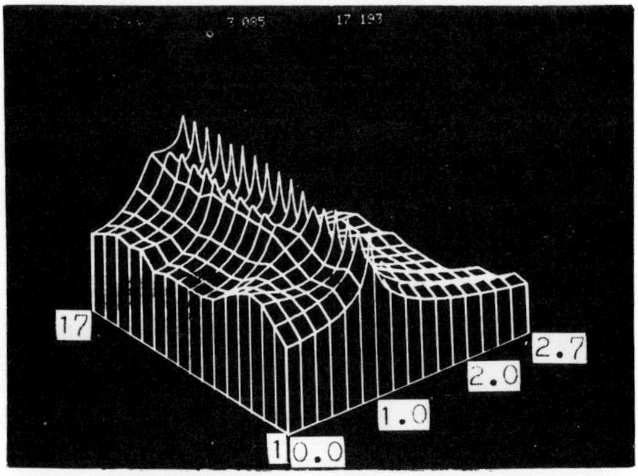

b) Burg's spectra in 3-D plot

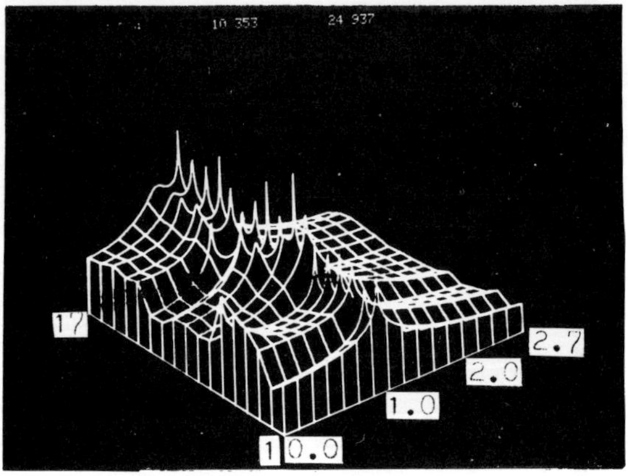

c) Fougere's spectra in 3-D plot

Figure 17

points are used for computation in frequency and filter order axes even though only 225 points are actually used for the display.

Figure 16a is a section of 25 data points from Fig. 15a to cover the period 1954-1978. Figure 16b shows the Burg's spectra for filter orders 1 to 17. The dynamic range is A=6.281, B=14.424, C=8.143. Figure 16c gives the corresponding spectra of the nonlinear method for filter orders of 1 to 17. The dynamic range is A=4.403, B=17.225 and C=12.822. The increase in dynamic range by using the nonlinear method is about 46.8 dB.

Next the spectral display of sinewaves will be considered. Fig. 17a is the sinewave examined which is given by

$$x(t) = \sin(2\pi ft + \theta) + \text{SNR}\ n(t)$$

with sampling period=0.05, f=1, θ=45, SNR=50 dB and n(t) is a zero mean unit variance Gaussian noise. Twenty-five points were used in the spectral analysis. Figure 17b shows the Burg's spectra for filter orders of 1 to 17 and frequency interval of 0 to 2.7 Hz. The dynamic range is A=-14.108, B=3.085, C=17.193. Figure 17c shows the spectra of the nonlinear method for the same filter orders and frequency interval. The dynamic range is A=-14.584, B=10.353, C=24.937. The increase in the dynamic range of the spectral magnitude by using the nonlinear method is about 77.5 dB.

REFERENCES

1. Fougere, P. F., "Spontaneous line splitting in Burg maximum entropy power spectra: a review," Proc. of the Second International Symposium on Computer-aided Seismic Analysis and Discrimination, N. Dartmouth, MA, Aug. 1981.
2. Chen., C. H., "A non-linear maximum entropy method for spectral estimation," Contract F49620-79-C-0038, Aug. 17, 1979.
3. Chen, C. H., J. Chen and C. Yen, "A minicomputer implementation of Fougere's maximum entropy spectrum analysis method," Technical Report, SMU-EE-TR-80-7, SMU, N. Dartmouth, MA, Aug. 20, 1980.
4. Chen, C. H., "Digital Waveform Processing and Recognition," CRC Press, Boca Raton, FL, Jan. 1982, Chapter 8.
5. Proceedings of the RADC Spectrum Estimation Workshop, Part II Problems and Solutions, Oct. 1979.

CHAPTER 4
Signal Classification, Detection and Resolution

1. Feature Selection for Signal Classification

For classification of the time series patterns, features can be extracted from the signal models as well as the spectral domain. The nonlinear methods provide accurate computation of the autoregressive signal model coefficients. For the coefficients of the autoregressive moving-average signal model, the computation is more involved. A nonlinear method has been suggested [1] to determine such coefficients. Another procedure [2] is to use spectral matching. This involves fitting an autoregressive moving-average model with a predetermined autoregressive part to the signal spectrum determined by the modified periodogram method. The signal models, when properly determined, allow for a considerable feature-space compression of the time series patterns. Two independent experimental studies [3][4] concerning teleseismic signal classification using autoregressive model coefficients as features, indicate that the model coefficients are not very effective for classification. Only 75% correct classification is achievable. Features derived from the accurately determined power spectrum give much better classification results.

Figure 1 shows the maximum entropy spectra of two explosion and two earthquake events. In the upper portion of each photograph is a plot of the digitized seismic waveform with 1200 data points at 10 samples/sec. The lower portion of the photograph is the maximum extropy power spectrum for 200 data points from the sample number 701 to 900. A spectral ratio is defined as the ratio of signal power above 0.5 Hz to that below 0.5 Hz. For the explosion event, the ratio tends to be greater than 1 while for the earthquake event, such ratio tends to be less than 1. The complete data set has 323 seismic samples (records) including 211 earthquake and 112 explosion events. For each class (event), 40 learning samples are used. Thus a total of 243 samples which do not include learning samples has been classified using the nearest neighbor decision rule. The spectral ratio for 200 data points from the sample number 550 to 749 and the spectral ratio for 200 data points for the sample number 750 to 949 are employed as two features in classification. The final classification result is 14 errors out of 243 or 94.24% correct recognition. This result is

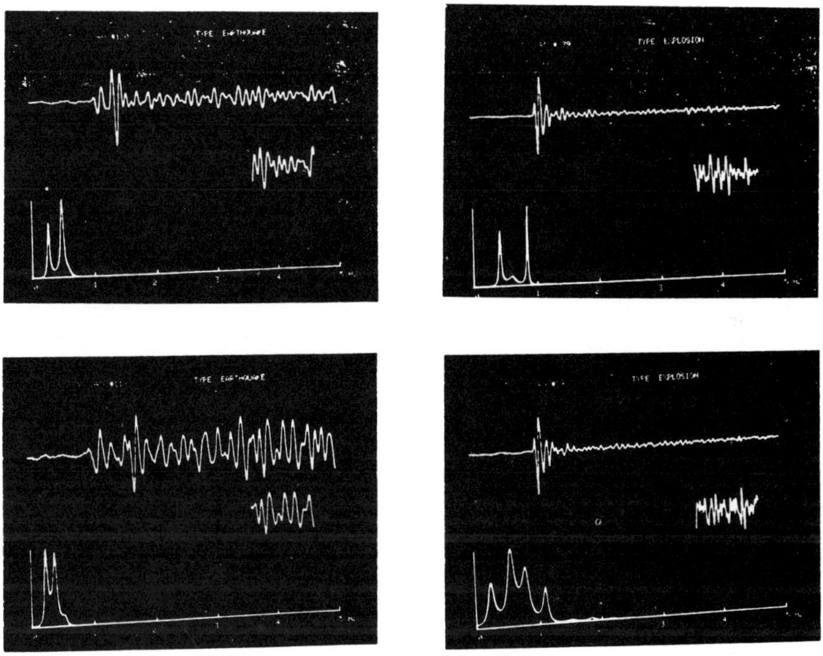

a) Spectra of 2 earthquake events   b) Spectra of 2 explosion events

Figure 1 The maximum entropy spectra of teleseismic events.

slightly better than the one obtained using a different set of the spectral features [5] computed from the modified periodograms. An interpretation of the better performance is that the more accurate spectrum obtained by the maximum entropy spectrum analysis provides more reliable geophysical discriminant for classification.

The spectral features which perform effectively as described above are empirically determined. Theoretical work on deriving the optimum spectral features remains to be done. The signal model parameters are designed to fit the data and thus may not be suitable for discriminating between different classes. However, if patterns of different classes can be adequately characterized by autoregressive moving-average models with distinct parameters, then these parameters can be used as features, yielding very high information compression rate. Such features are likely to facilitate good discrimination [4]. Since this is not usually possible for the real data, new signal models must be examined that emphasize the differences among various classes. Again further work is needed in this direction.

2. Detection of Signals in Noise

The signal detection capability can be greatly increased with the use of the nonlinear maximum entropy spectral analysis. The signals are either sinusoidal or consisting of certain periodicities (see e.g.[6]) that must be detected in the presence of additive noise. The maximum entropy spectral analysis tends to enhance the periodic components of the data making them easier to detect. The experimental comparisons of the (Fourier) periodogram method and the maximum entropy method have consistently demonstrated that for a given signal-to-noise ratio, the detection performance with the maximum entropy spectral analysis processing is far superior to the Fourier spectrum analysis. For a given probability of false alarm, the probability of detection is usually related to the signal-to-noise ratio after spectral processing by a Gaussian integral. However, the signal-to-noise ratio relationship before and after the maximum entropy spectral analysis has not yet been established. Thus it is not possible to predict theoretically the performance of the detection scheme employing the maximum entropy spectral analysis processing. For multiple signals in noise, the detection performance for each signal may not be the same because of the nonlinear method used for the spectral analysis. In the case when signal tracking is required, in addition to detection, the available data segment for processing at a given instant is likely to be very short. In such a situation the maximum entropy spectral analysis will give better results than classical methods, provided that the computation required can be done in real time.

3. Signal Resolution

To determine the capability of resolving two or more tonal signals in noise is another way to test the spectral analysis methods. Recently, there has been an active research interest in determining and comparing the spectral resolution capability of the maximum entropy spectral analysis and related methods (see e.g. Wenzel and Zürn [7]) by using sinewaves. Let Δf be the frequency separation of two sinewaves and T be the signal duration considered. For classical (Fourier) techniques, the product $(\Delta f)T$ is greater than 1. Numerous experiments have shown that the product is less than 1 with the maximum entropy spectral analysis method, indicating the superior resolving power of the method. Using synthetic signals consisting of several decaying sinusoids with random noise superimposed, Wenzel and Zürn have found the approximate limit of resolution to correspond to $(\Delta f)T = 0.2$; but this is not a well-defined limit.

Among the related methods of data-adaptive spectral analysis is the maximum likelihood spectral estimate. The maximum likelihood estimate is a minimum variance unbiased estimator of the spectrum and may be derived by solving a classical optimum filtering problem [8]. The inverse transform of the maximum likelihood estimate does not in general agree with the measured autocorrelation values. The maximum likelihood estimate is in fact an average of various maximum entropy spectral estimates [8]. This implies that the maximum likelihood estimate is more stable statistically but has less spectral resolution than the maximum entropy estimate. Recently a new high resolution frequency estimation method has been proposed by Tufts and Kumaresan [9]. A careful examination of the spectral resolution capability of the (Fougere) nonlinear method has demonstrated [10] that the nonlinear method is better by using identical data employed in [9]. Figure 2a shows the 25-point discrete cosine and sinewaves containing frequency components of 0.5 Hz and 0.52 Hz with a sampling rate of 1 sample/sec. A complex sinusoid is formed by these two waves as

$$y(n) = \exp j(2\pi f_1 n + \theta) + \exp j(2\pi f_2 n + \theta) + w(n); \qquad n=1,2,\ldots,N \tag{1}$$

where $\theta=45^\circ$, $N=25$, and $w(n)$ is the complex Gaussian noise. The spectral plots (Fig. 2b) of the complex sinusoid with 10 filter weights clearly demonstrate that, at the 5 dB signal-to-noise ratio, the nonlinear method (solid curve) can resolve the two frequency components, while the Burg's method (dotted curve) cannot. The frequency shifting that appears in [9] is completely removed in the nonlinear method. The product of frequency separation and signal duration is 0.5 for both References 9 and 10. For the real sinewave case which is not reported in [9], the frequency separation must be larger than 0.02Hz. Figure 2c shows that for 0.5 Hz and 0.54 Hz frequency components, the nonlinear method can resolve better than the Burg's method.

Another performance comparison is to use the variance of the frequency estimate as a criterion. The straight line shown in Figure 3 is the Cramer-Rao bound for the maximum likelihood estimate as determined from [11]. For the same filter order of 10, the nonlinear method has a much lower "threshold signal-to-noise ratio" of 8.5 dB as compared to 10 dB reported in [9]. The better resolution capability of the nonlinear method will provide a better signal detection performance as suggested in [12]. Note that the Cramer-Rao bound is the ideal limiting performance (as $N\to\infty$) not achievable with a finite number of data points.

As a final remark of this Chapter, it should be pointed out that the data adaptive capability of the maximum entropy method depends critically on the order of the prediction error filter. The best signal resolution, signal detection and classification can be obtained

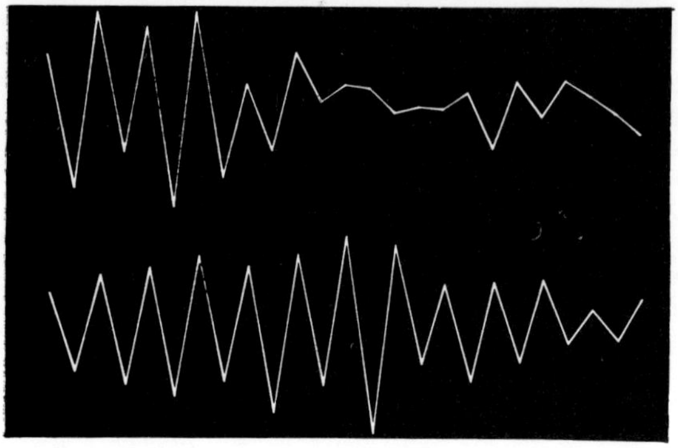

a) Real(upper) and imaginary(lower) parts of a 25-point two-frequency complex sinusoids plus Gaussian noise

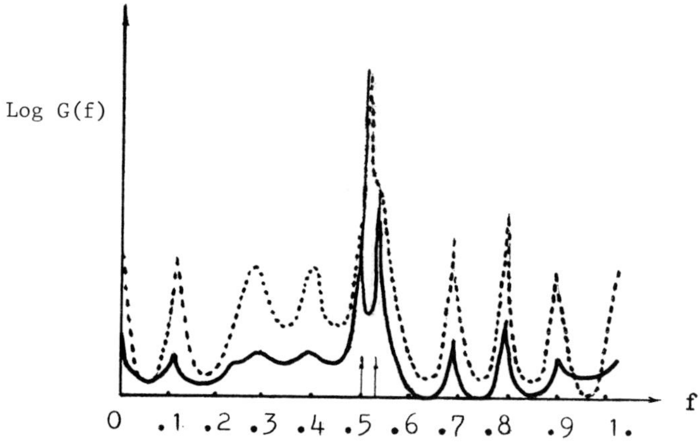

b) Corresponding Fougere's spectrum(solid curve) and Burg's spectrum(dotted curve) at SNR=5dB

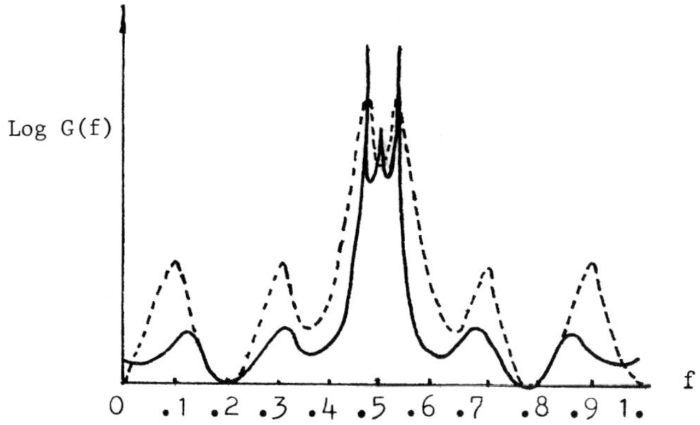

c) Fougere's spectrum(solid curve) and Burg's spectrum(dotted curve) for two-frequency real sinewaves plus Gaussian noise at SNR=5dB

Figure 2

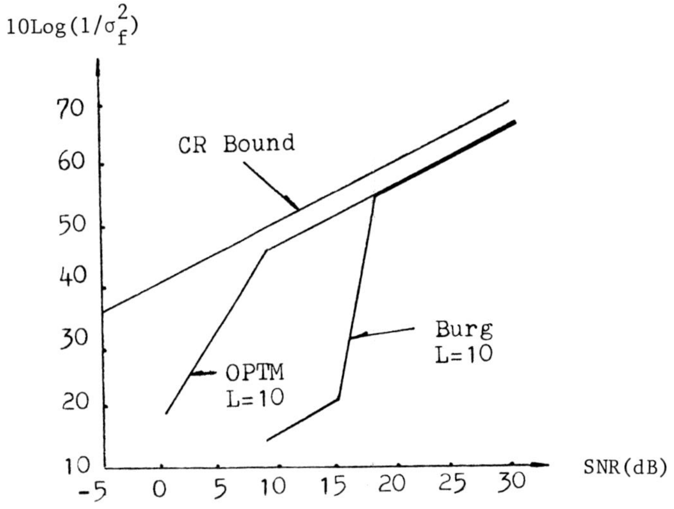

Figure 3 Performance comparison of frequency estimation methods. OPTM is the Fougere's method.

with a proper choice of filter order. Our experience along with the Akaike criterion suggests that the proper filter order should be one-third to one-half of the number of data points used.

REFERENCES

1. Kittler, J., "A nonlinear method of feature selection for stationary pattern vectors," Proc. of the Third International Joint Conference on Pattern Recognition, pp. 64-67, 1979.
2. Chen, C. H., "Digital Waveform Processing and Recognition," CRC Press, Inc., Boca Raton, FL, Jan. 1982, Chapter 7.
3. Kwan, T. W. F., "Spectral analysis of the seismic data," Master thesis, SMU E. E. Dept., N. Dartmouth, MA, January 1979.
4. Sarna, C. S. and H. Stark, "Pattern recognition of waveforms using autoregressive modeling with application to earthquake/explosion data," Proc. of the Second International Symposium on Computer-aided Seismic Analysis and Discrimination, pp. 138-143, Aug. 1981.
5. Chen, C. H., "Seismic pattern recognition," Geoexploration journal, vol. 16, pp. 133-146, 1978.
6. Berni, A. J., "Target identification by natural resonance estimation", IEEE Trans. on Aerospace and Electronic Systems, vol. AES-11, No. 2, pp. 147-154, March 1975.
7. Wenzel, F. and W. Zürn, "Numerical investigation of the spectral resolving power of Burg's maximum entropy method," Journal of Geophysics, vol. 48, pp. 121-123, 1980.
8. Childers, D. G. (editor), "Modern Spectrum Analysis", IEEE Press, New York 1978; also distributed by John Wiley & Sons, New York.
9. Kumaresan, R. and D. W. Tufts, "Improved spectral resolution III: Efficient realization," Proc. of the IEEE, vol. 68, no. 10, pp. 1354-1355, October 1980.
10. Chen, C. H., "Spectral resolution of Fougere's maximum entropy spectral analysis," Proc. of the IEEE, vol. 69, no. 7, pp. 839-841, July 1981.
11. Rife, D. C., "Digital tone parameter estimation in the presence of Gaussian noise," Ph.D. dissertation, Polytechnic Institute of Brooklyn, June 1973.
12. Dyson, T. and S. Rao, "Some detection and resolution properties of maximum entropy spectral analysis," Signal Processing Journal, vol. 2, pp. 261-270, 1980.

CHAPTER 5

On Multichannel (Multivariate) Maximum Entropy Spectral Analysis

1. Introduction

The univariate maximum entropy spectral analysis is now well developed and it has been successfully applied to many research areas involving radar, sonar and geophysical data. There has been some work done to extend the maximum entropy spectral analysis to the multivariate case. Whittle [1] and Robinson [2][3] generalized the Levinson-Durbin recursion to the multivariate case by fitting both the forward and backward autoregressions in a stepwise fashion. In his thesis, Burg [4] mentioned the multichannel case. However, the computer programs for both multichannel and multivariate maximum entropy spectral analysis have only recently been developed successfully. Morf, et.al [5] developed an algorithm for direct estimation of the normalized reflection coefficients from the observed data for maximum entropy spectral analysis. They also compared this spectral estimation approach with the methods of Jones [6], Nuttall [7] and Strand [8], which are more direct extensions of Burg's work to the multichannel (multivariate) case. Burg's algorithm does not generalize directly since the forward and backward autoregression matrices are not the same. Also, the forward and backward one-step prediction error covariance matrices are different, although they have the same determinant. In this Chapter, we shall use the generalization method of Strand and apply it to multichannel data and imagery data in addition to a set of test signals. In spite of the programming complexity, the multichannel and multivariate maximum entropy spectral analysis will have increased application potential as real data is almost always gathered in several channels simultaneously. Data from several channels form a vector for the multivariate study.

Although for short multichannel real sinusoids, the spectral line shifting and spontaneous line splitting are observed [9], we have demonstrated that such phenomena are almost completely eliminated in the complex signal case. This result is practically significant as it indicates that accurate spectral estimation is possible for the complex signals. Good phase estimates can be obtained from the multichannel maximum entropy spectral analysis, from which time delays among spatially separated sensors can be properly determined.

Two sets of computer programs have been prepared and extensively tested. Appendix D is the long version based on the original software developed by Strand [10]for complex signals. It has been operated with our DEC-20 Computer. Appendix E is the short version, for real signal only, adapted from the work of Strand. This version has been operated with our PDP 11/45 minicomputer. Both programs perform almost identically for short multichannel real sinewaves as a test signal.

2. Theoretical Background

Let

$$y(t) = \begin{bmatrix} y_1(t) \\ y_2(t) \\ \cdot \\ \cdot \\ \cdot \\ y_p(t) \end{bmatrix} \quad ; \quad t = \{nT\}_{n=-\infty}^{\infty} \tag{1}$$

be a complex zero-mean, wide-sense stationary p-channel time series, where T is the sampling period. The N-element forward filter (or forward filter of length N) has the form,

$$y(t) + \sum_{k=1}^{N} F_{kN}^{*} y(t-kT) = e_N(t) \tag{2}$$

Similarly, we consider the N-element backward filter

$$y(t) + \sum_{k=1}^{N} B_{kN}^{*} y(t+kT) = b_N(t) \tag{3}$$

In Eqs. (2) and (3), the forward and backward filter coefficients, F_{kN} and B_{kN}, respectively, are complex pxp matrices. The asterisk * denotes the (Hermitian) conjugate transpose. For the optimum forward or backward filter Eq. (2) or Eq. (3), we impose the condition that the expected mean-square value of $e_N(t)$ or $b_N(t)$ should be a minimum. For Eq. (2) this involves determining F_{1N}, F_{2N}, $\cdots$, F_{NN} such that

$$E\{e_N^{*}(t)e_N(t)\} = \text{minimum} \tag{4}$$

where $E\{\cdot\}$ denotes the expected value. Inserting Eq. (2) into Eq. (4) and minimizing with respect to the coefficients $F_{kN}{*}$ gives the system

$$\begin{bmatrix} R_o & R_1 & R_2 & \cdots\cdots R_N \\ R_{-1} & R_o & R_1 & \cdots\cdots R_{N-1} \\ \cdot & \cdot & \cdot & \cdot \\ \cdot & \cdot & \cdot & \cdot \\ \cdot & \cdot & \cdot & \cdot \\ R_{-N} & R_{-N+1} & R_{-N+2} & R_o \end{bmatrix} \begin{bmatrix} I \\ F_{1N} \\ \cdot \\ \cdot \\ \cdot \\ F_{NN} \end{bmatrix} = \begin{bmatrix} P_N \\ 0 \\ \cdot \\ \cdot \\ \cdot \\ 0 \end{bmatrix} \tag{5}$$

where I is the pxp identity matrix and the square block submatrices in this system are defined by

$$R_k = E\{y(t)\ y^*(t-kT)\} \quad , k=0,1,\ldots N \tag{6}$$

so that

$$R_{-k} = R_k^* \tag{7}$$

We call the coefficient matrix in Eq. (5) the R-matrix, or simply R. The forward power matrix P_N for the resulting optimum filter (that is, for F_{kN} satisfying Eq. (5)) is

$$P_N = E\{e_N(t)\ e_N^*(t)\} \tag{8}$$

By multiplying both sides of Eq. (5) by F_N^* we have,

$$P_N = F_N^*\ R\ F_N \tag{9}$$

where we define F_N by

$$F_N^* = [I, F_{1N}^*, \ldots, F_{NN}^*]$$

with comma indicating matrix partition.

Similarly, for the optimum backward filter, we can derive a system:

$$\begin{bmatrix} R_o & R_{-1} & R_{-2} \cdots & R_{-N} \\ R_1 & R_o & R_{-1} \cdots & R_{-N+1} \\ \vdots & \vdots & \vdots & \vdots \\ R_N & R_{N-1} & R_{N-2} \cdots & R_o \end{bmatrix} \begin{bmatrix} I \\ B_{1N} \\ \vdots \\ B_{NN} \end{bmatrix} = \begin{bmatrix} P'_N \\ 0 \\ \vdots \\ 0 \end{bmatrix} \tag{10}$$

We call the coefficient matrix in Eq. (10) the R' matrix, or simply R'. The backward power matrix P'_N for the resulting optimum filter is

$$P'_N = E\{b_N(t)\ b_N^*(t)\} \tag{11}$$

and it follows from Eq. (10) that

$$P'_N = B_N^*\ R'\ B_N \tag{12}$$

where

$$B_N^* = [I, B_{1N}^*, \ldots, B_{NN}^*]$$

Burg has derived the following results concerning the recursive solution of Eq. (5) and Eq. (10), the forward and backward filters may be postulated in the form

$$F_N = \begin{bmatrix} F_{N-1} \\ 0 \end{bmatrix} + \begin{bmatrix} 0 \\ B'_{N-1} \end{bmatrix} C_N \qquad \text{and}$$

$$B_N = \begin{bmatrix} F_{N-1} \\ 0 \end{bmatrix} C_N' + \begin{bmatrix} 0 \\ B_{N-1}' \end{bmatrix} \tag{13}$$

where $B_{N-1}' = [B_{N-1,N-1}^*, \ldots, B_{1N}^*, I]^*$ is the block reverse of B_{N-1}, and the pxp matrices C_N and C_N' are called forward and backward reflection coefficients, respectively. It can be shown that P_N and P_N' may be computed by using the modern Levinson algorithm [4],

$$P_N = P_{N-1} - C_N^* P_{N-1}' C_N \tag{14}$$

$$P_N' = P_{N-1}' - (C_N')^* P_{N-1} C_N' \tag{15}$$

and

$$C_N = (P_{N-1}')^{-1} (C_N')^* P_{N-1}$$

and

$$C_N' = (P_{N-1})^{-1} C_N^* P_{N-1}' \tag{16}$$

For the positive definite $P_N > 0$ (and $P_N' > 0$), it is known that the maximum entropy (autoregressive) power spectral density, S(f), a pxp matrix, can be calculated from the forward filter F_N by

$$S(f) = T[F^{-1}(z)]^* P_N [F^{-1}(z)] \tag{17}$$

where $F(z) = I + F_{1N}z + \ldots + F_{NN}z^N$, z is the complex scalar defined by $z = \exp(2\pi jfT)$ and f is the frequency.

It can be shown (as suggested in Eq. (9)) that $P_o, P_1, \ldots, P_N$ and $(P_o', P_1', \ldots, P_N')$ are positive definite if and only if R is positive definite. Instead of trying to solve Eq. (5) using approximations to the R_k, Strand [8] derived a method of estimating the reflection coefficients C_N and C_N' by minimizing a certain weighted sum of squares of residuals for the forward and backward filters, after which all forward and backward filter coefficients F_{kN} and B_{kN} are obtained by the modern Levinson algorithm. It can be shown that

$$B\, C_N + P_{N-1}' C_N P_{N-1}' E = -2G \tag{18}$$

where

$$E = \sum_{m=1}^{M} W_m e_m^N (e_m^N)^*$$

$$G = \sum_{m=1}^{M} W_m b_m^N (e_m^N)^* \tag{19}$$

$$B = \sum_{m=1}^{M} W_m b_m^N (b_m^N)^*$$

and W_m are positive scalar weights (usually all taken as 1/M in practice). Here N_d is the number of observations of a realization of y(t), and $M = N_d - N$.

Also e_m^N and b_m^N can be obtained from

$$e_m^N = e_{m+1}^{N-1} + C_{N-1}^* \, b_{m+1}^{N-1}$$

and

$$b_m^N = b_m^{N-1} + (C_{N-1}')^* \, e_m^{N-1} \tag{20}$$

$$m=1,2,\ldots N_d-N = M \;,\; N > 1$$

The whole computing procedure is illustrated in Fig. 1, where $y_m = y(t=mT)$.

$$P_0 = P_0' = \left(\sum_{m=1}^{N_d} y_m \, y_m^* \right)/N_d$$

Let $e_m' = y_{m+1}$ and $b_m' = y_m$, $m=1,2,\ldots N_d-1$

↓

Compute E, B, G by (19)

↓

Solve C_N and C_N' by (18) and (16) respectively.

↓

Compute P_N and P_N' by (14) and (15) respectively.

↓

Update F_N and B_N by (13)

↓

Obtain new e_m^N and b_m^N by (20) (loop back to Compute E, B, G)

↓

Calculate spectra by (17)

FIG. 1 Computing Procedure for the Multichannel Maximum Entropy Spectral Analysis

3. Numerical Results for a Test Signal

To ensure the correctness of the program, the same data which Strand used have been tested:

N_d (number of data points sampled) = 128

T = 1 second.

p (number of channels) = 2

$$y_t = \begin{bmatrix} y_{1t} \\ y_{2t} \end{bmatrix} \;,\quad t=1,2,\ldots,\quad 128$$

$$\text{Re } y_{1t} = \cos(2\pi f_i t) + 0.25 (\text{Ran f} - 0.5)$$

$$\text{Im } y_{1t} = \sin(2\pi f_i t) + 0.25 (\text{Ran f} - 0.5) \quad (21)$$

$$\text{Re } y_{2t} = \cos(2\pi f_i t + \phi_d) + 0.25 (\text{Ran f} - 0.5)$$

$$\text{Im } y_{2t} = \sin(2\pi f_i t + \phi_d) + 0.25 (\text{Ran f} - 0.5)$$

where Ranf is a pseudo-random number uniformly distributed between 0 and 1, ϕ_d is the phase difference between channels and f_i is the input frequency. If we denote the spectrum by

$$S(f) = \begin{bmatrix} S_{11}(f) & S_{12}(f) \\ S_{21}(f) & S_{22}(f) \end{bmatrix} \quad (22)$$

where f is the frequency in Hz, then the squared coherence is given by

$$\text{coh}^2(f) = \frac{|S_{12}(f)|^2}{S_{11}(f)\, S_{22}(f)} \quad (23)$$

A rough indication of the expected performance is given by Akaike's FPE criterion, given by

$$\text{FPE}(N) = \det P_N \left(\frac{N_d + 1 + pN}{N_d - 1 - pN}\right)^p$$

Figs. 2a and 2b show $S_{11}(f)$ and $S_{22}(f)$ for the real sinusoid inputs with f = 0.0625 Hz, ϕ_d = 1 rad. and NPEF (number of terms in the prediction error filter) = 7. The results for $S_{12}(f)$ are shown (in amplitude and phase) in Figs. 2c and 2d. From Fig. 2d, we can observe that the phase of $S_{12}(f)$ is equal to -1 radian at f = 0.0625 Hz. In Fig. 2e, the squared coherence is shown. Fig. 2f and Fig. 2g show the maximum and minimum eigenvalues of spectra. Figs. 3a-3g are based on the repeated procedure of Figs. 2a-2g but with complex input signals. The comparison shows that the complex multichannel maximum entropy method gives better resolution and accuracy than the real case.

It is well known that the phase difference between channels is a very important parameter. Correct phase difference can give a very good estimation of time delay so as to detect the location of the target. Different sets of data have been tested using the complex multichannel maximum entropy program with phase differences ϕ_d = 0.5 rad, -1.rad,2. rad. (f_i=0.0625Hz), and ϕ_d = -0.5 rad (f_i=0.2 Hz). The results of 1-1 spectrum $S_{11}(f)$ and phase of 1-2 spectrum, arg. $S_{12}(f)$ are shown in Figs. 4(a,b), 5(a,b), 6(a,b) and 7(a,b) respectively. It is noted that the results are satisfactory.

4. The Problem of Line Splitting and Frequency Shifting

In order to study this problem we have tested a signal which exemplified the problem for the single-channel signals but in both real and

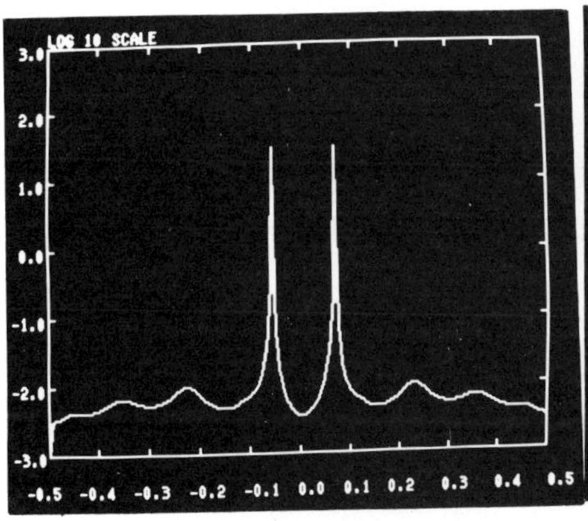

a) $S_{11}(f)$

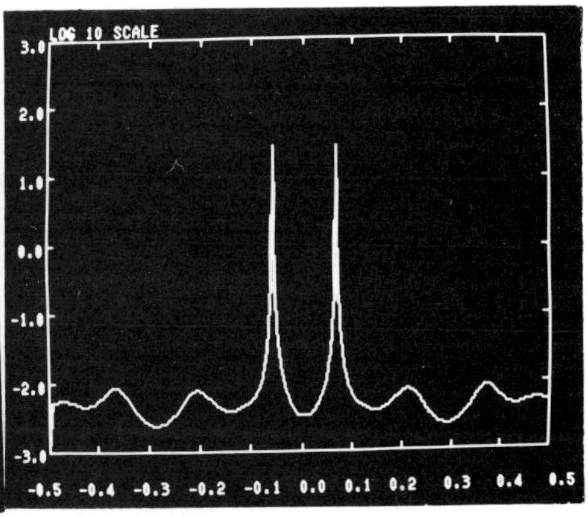

b) $S_{22}(f)$

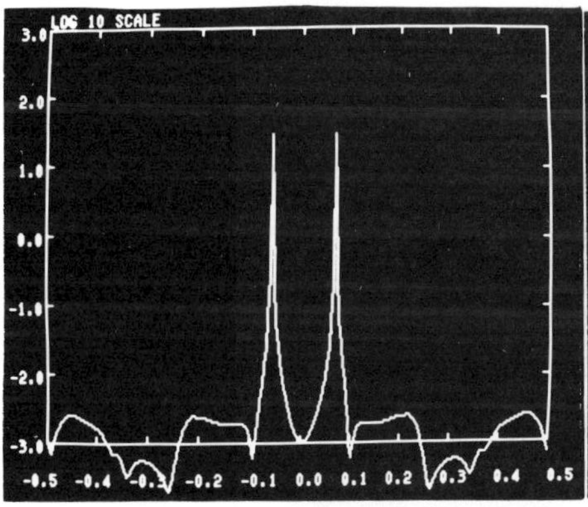

c) Amplitude of $S_{12}(f)$

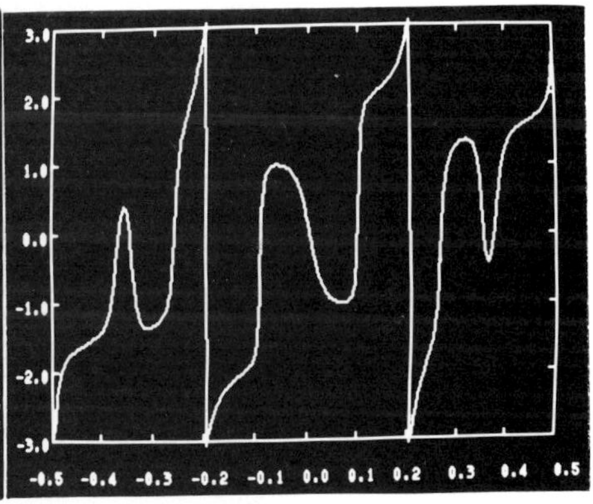

d) Phase of $S_{12}(f)$

Figure 2 Spectral analysis for two-channel real signals.

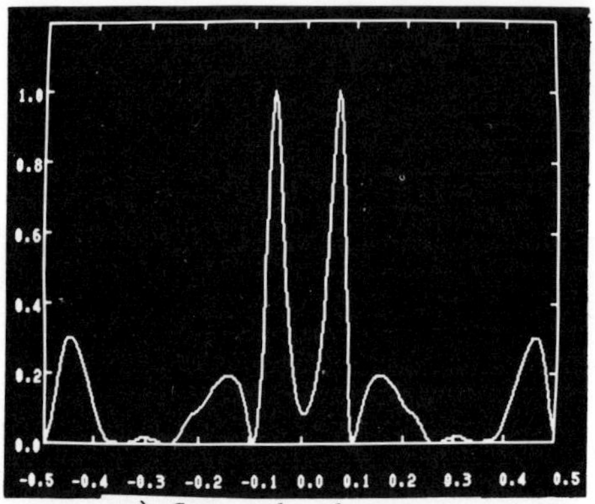

e) Squared coherence

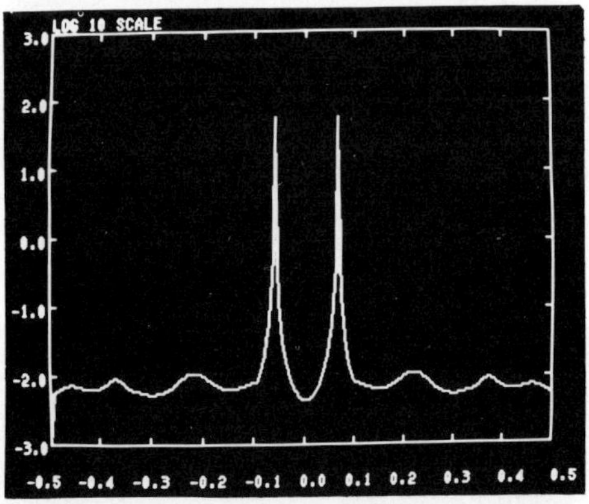

f) Maximum eigenvalue of the spectrum

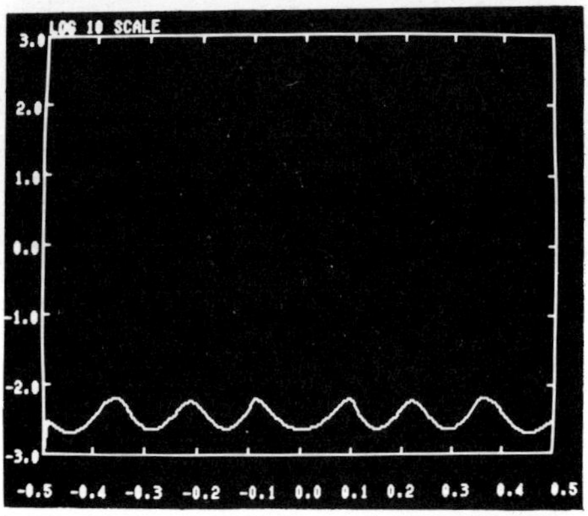

g) Minimum eigenvalue of the spectrum

Figure 2 (continued)

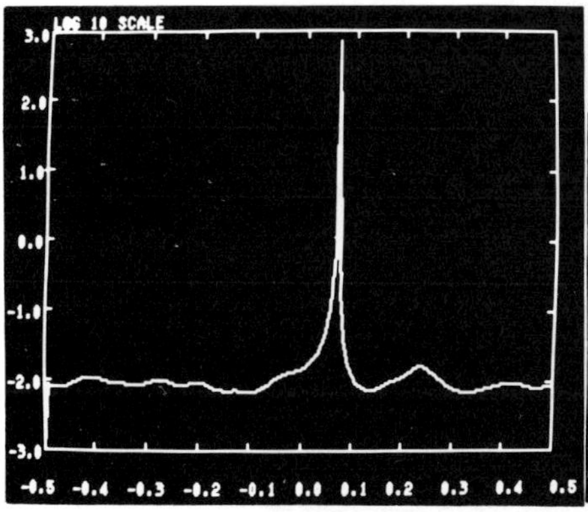

a) $S_{11}(f)$

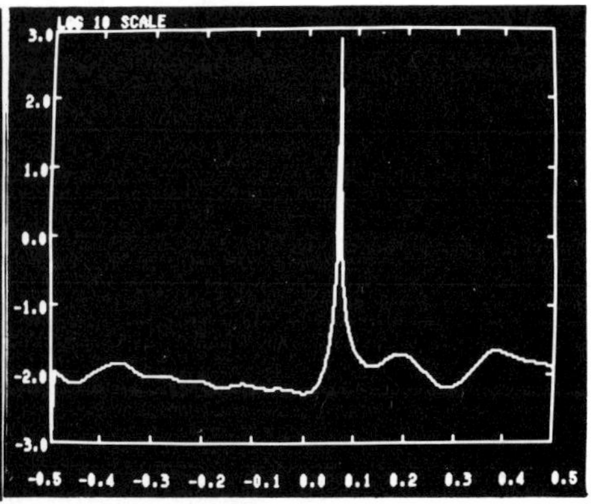

b) $S_{22}(f)$

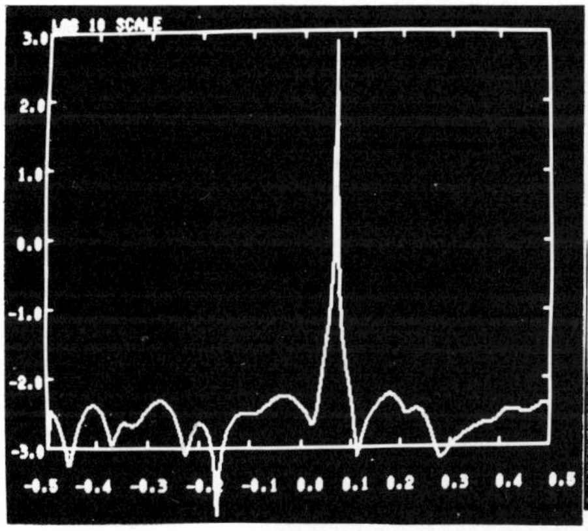

c) Amplitude of $S_{12}(f)$

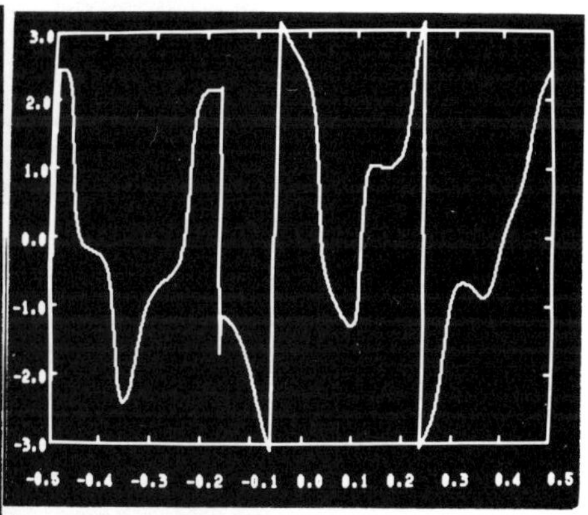

d) Phase of $S_{12}(f)$

Figure 3 Spectral analysis for two-channel complex signals.

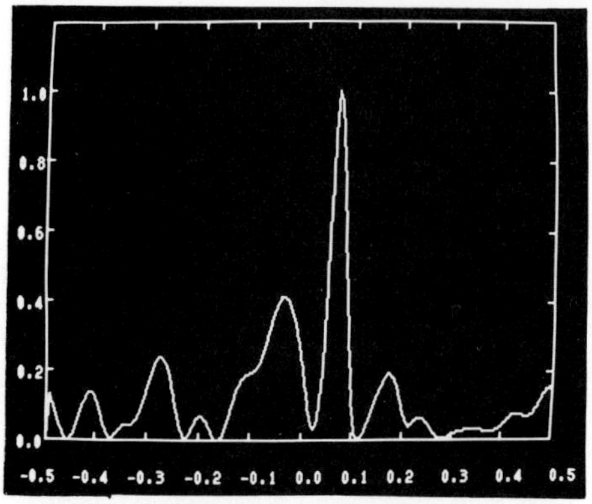

e) Squared coherence

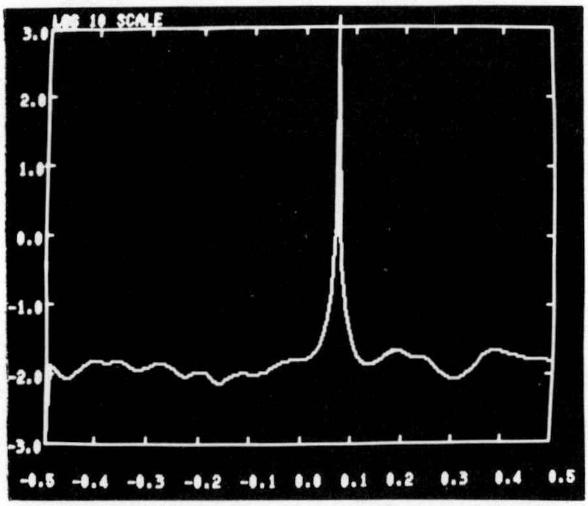

f) Maximum eigenvalue of the spectrum

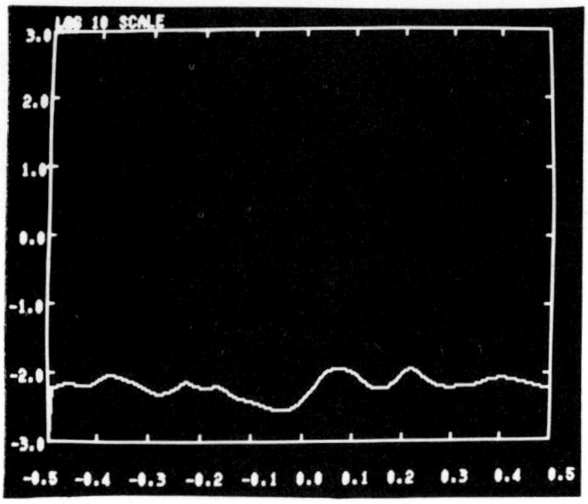

g) Minimum eigenvalue of the spectrum

Figure 3 (continued)

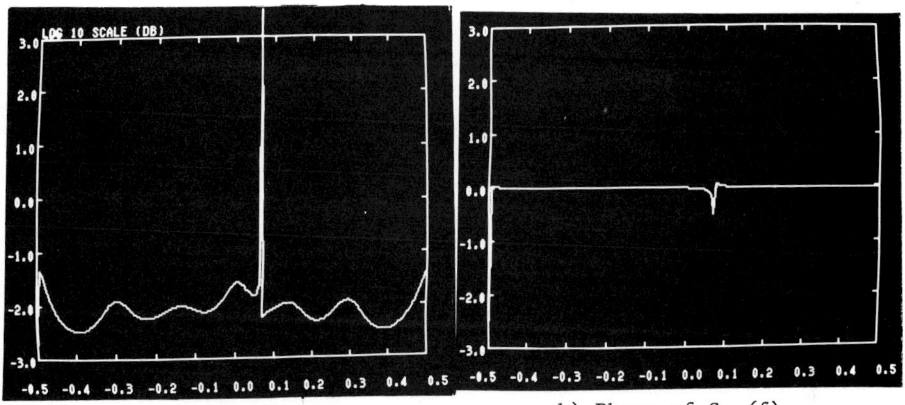

a) $S_{11}(f)$ b) Phase of $S_{12}(f)$

Figure 4 Two-channel spectra with ϕ_d=0.5 rad.

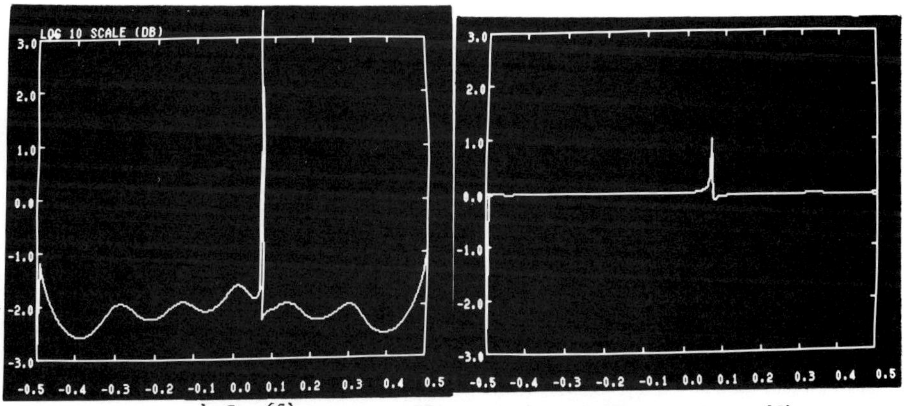

a) $S_{11}(f)$ b) Phase of $S_{12}(f)$

Figure 5 Two-channel spectra with ϕ_d=-1. rad.

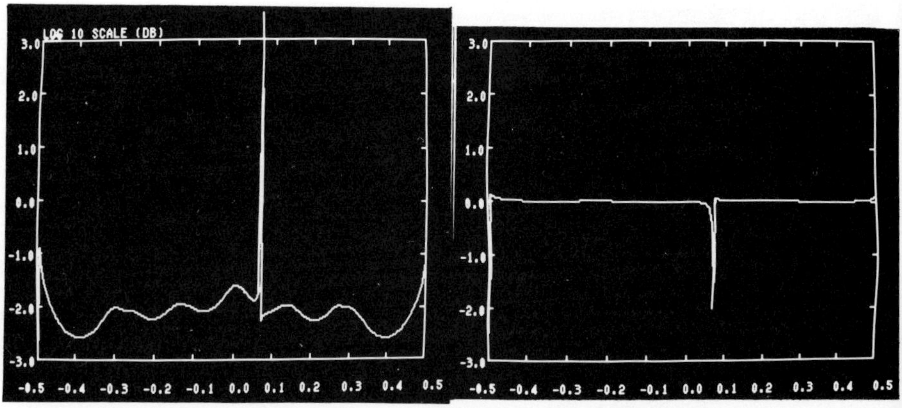

a) $S_{11}(f)$ b) Phase of $S_{12}(f)$

Figure 6 Two-channel spectra with ϕ_d=2. rad.

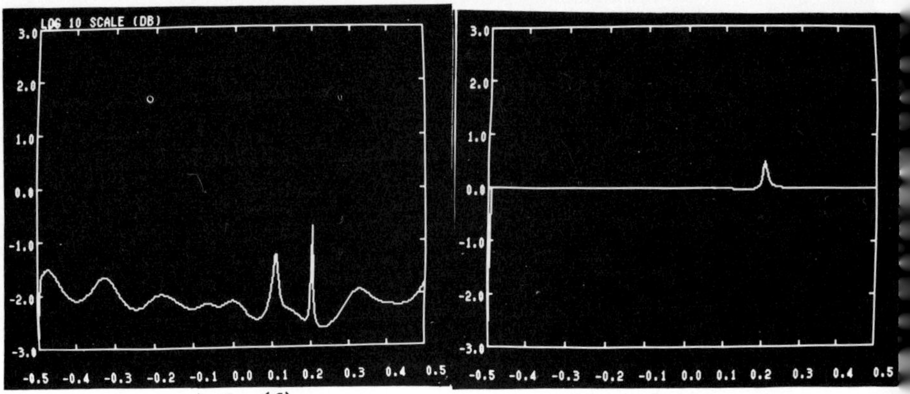

a) $S_{11}(f)$ b) Phase of $S_{12}(f)$

Figure 7 Two-channel spectra with ϕ_d=-0.5 rad and f_i=0.2 Hz.

complex cases. This is a single sinewave of frequency 1 Hz sampled every 0.05 seconds, with additive noise:

$$\mathrm{Re}\ f_1(n) = \cos(\ (n-1)\pi/10 + \phi_1) + Ag_1(n)$$

$$\mathrm{Im}\ f_1(n) = \sin(\ (n-1)\pi/10 + \phi_1) + Ag_1(n)$$

$$\mathrm{Re}\ f_2(n) = \cos(\ (n-1)\pi/10 + \phi_2) + Ag_2(n)$$

$$\mathrm{Im}\ f_2(n) = \sin(\ (n-1)\pi/10 + \phi_2) + Ag_2(n)$$

where ϕ_1 and ϕ_2 are initial phases, $g_1(n)$ and $g_2(n)$ are independent realizations of random numbers in the range ±0.5, and A is the noise amplitude. Also, we used N_d and NPEF to represent the number of data points and the number of terms in the prediction error filter respectively.

The number of terms in the prediction error filter (NPEF) affects the quality of the spectrum as illustrated by Fig. 8a with NPEF = 3,9,21. The same data are applied to the real and complex multichannel cases as shown in Fig. 8b, 8c, respectively. The real multichannel case is seen to be worse even than the single channel case (for NPEF = 21), but the complex signal performs well. Figs. 9-11 show the effects of the number of data samples (N_d) to the location of spectral peak of real signals. Figs. 9a, 10a, 11a show the location of spectral peak to be farther from 1 Hz as the cycles are not completed. When the number of data points increases, the frequency shift decreases as shown in Fig. 14. As expected there is a node at N_d = 21, 41,..., for every full cycle, where there is an excellent agreement between the true and apparent frequencies. Figs. 12a, 13a show two more cases but with different initial phase angles. The locations of spectral peak for various number of data are also shown in Figs. 15 and 16, respectively (for ϕ_1=45, ϕ_2=135 and ϕ_1=-45, ϕ_2=-135). Again we have noted the same property which is affected by the number of data points sampled. Comparing the location of the peak of N_d=21 of Figs. 9a, 12a and 13a, and the vertical axis of Figs. 14, 15, and 16, we have noted the peak of the real multichannel case is also affected by the initial phase. The peak with 45°, 135° phases shifts farther than with 0°, 180° case. This is the same condition as in the single channel case.

It is important to note that for multichannel complex signals the location of spectral peaks does not shift away from the true value for various variations (number of data sampled, number of terms in the prediction error filter, value of initial phase). From Figs. 9b-13b and Figs. 14-16, the complex multichannel case always gives very satisfactory results. Therefore, we can conclude that for complex signals, line splitting and spectrum shifting will not be a problem in multichannel maximum entropy spectral analysis.

5. Spectral Analysis of Geophysical Data

A three-channel geophysical data set provided by Robinson [11] includes sunspot numbers, northern light activity and earthquake activity index with 100 data points each, as shown in Fig. 17a. Fig. 17b is the spectral plot of each channel on a linear scale, while Fig. 17c

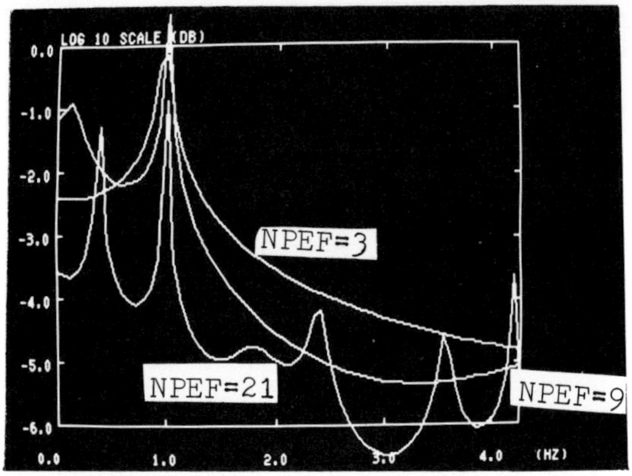

a) Maximum entropy power spectra of single channel real data (ϕ_1=0, A=0.05, N_d=24, NPEF=3,9,21)

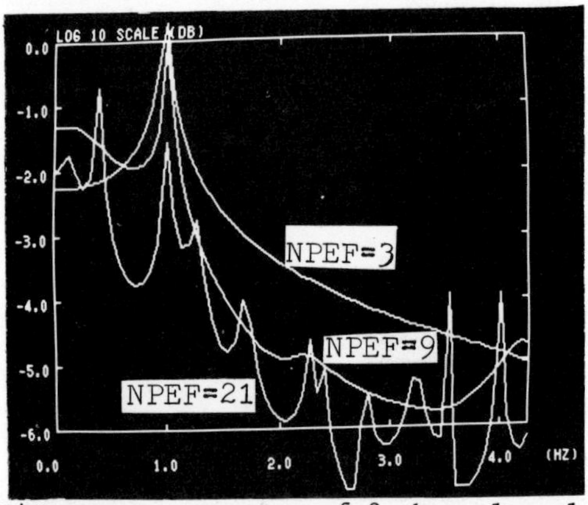

b) Maximum entropy power spectra of 2-channel real data (ϕ_1=0, ϕ_2=180,A=0.05, N_d=24, NPEF=3,9,21)

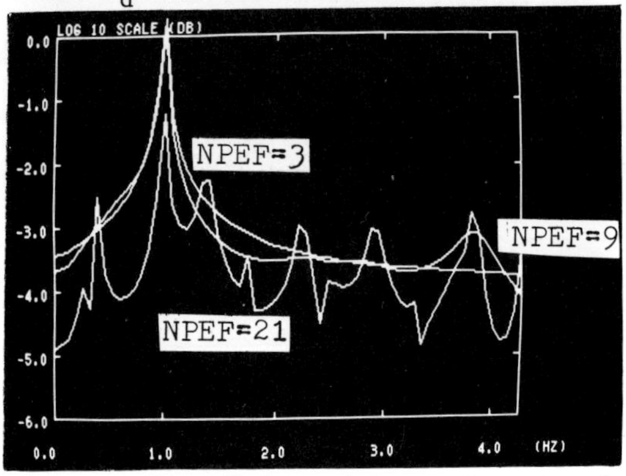

c) Same as Fig. 8b but with complex signals

Figure 8

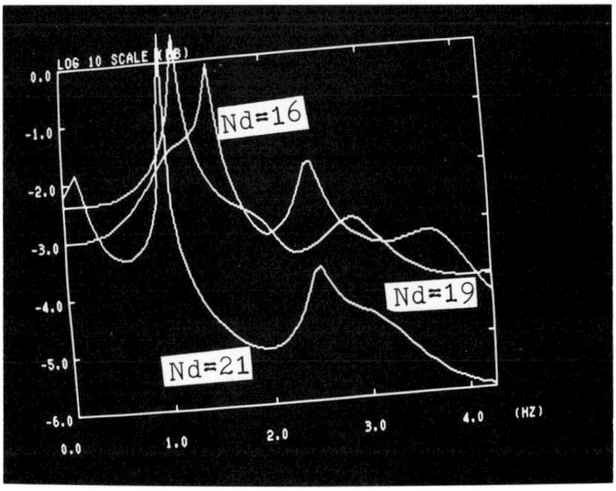

a) Maximum entropy power spectra for 2-channel real data (ϕ_1=0,ϕ_2=180, A=0.05, NPEF=9, N_d=16,19,21)

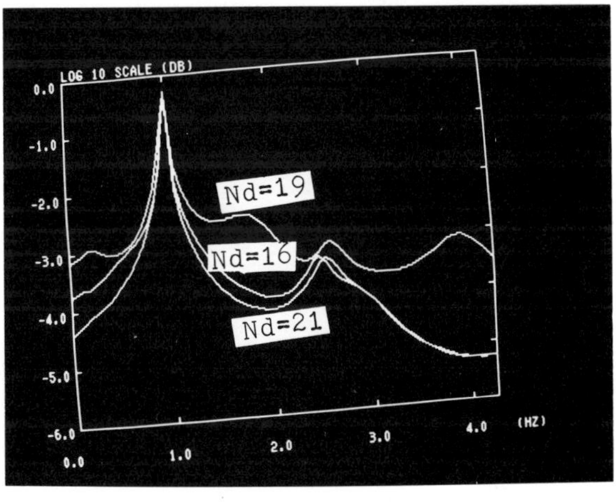

b) Same as Fig. 9a but with complex signals

Figure 9

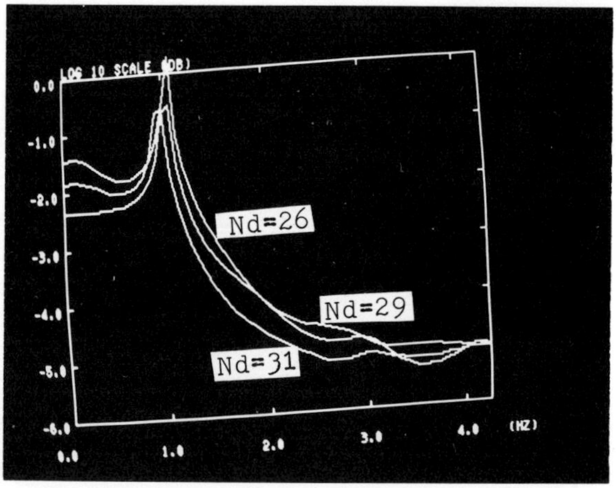

a) Maximum entropy power spectra for 2-channel real data (ϕ_1=0,ϕ_2=180, A=0.05, NPEF=9, N_d=26,29,31)

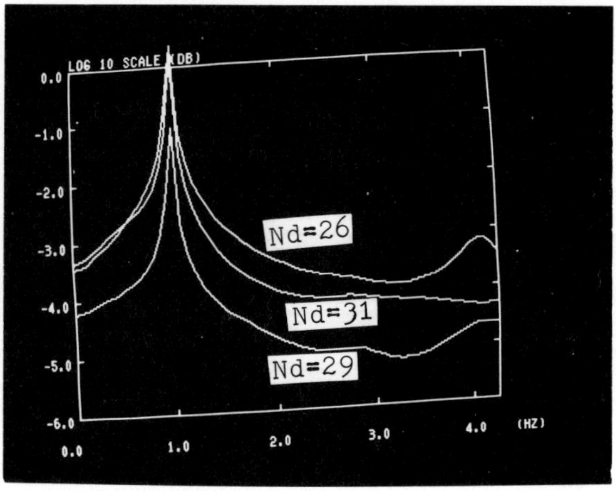

b) Same as Fig. 10a but with complex signals

Figure 10

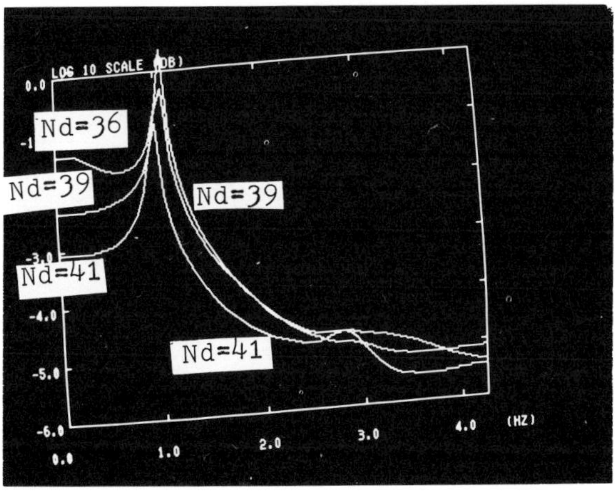

a) Maximum entropy power spectra of 2-channel real data (ϕ_1=0,ϕ_2=180, A=0.05, NPEF=9, N_d=36,39,41)

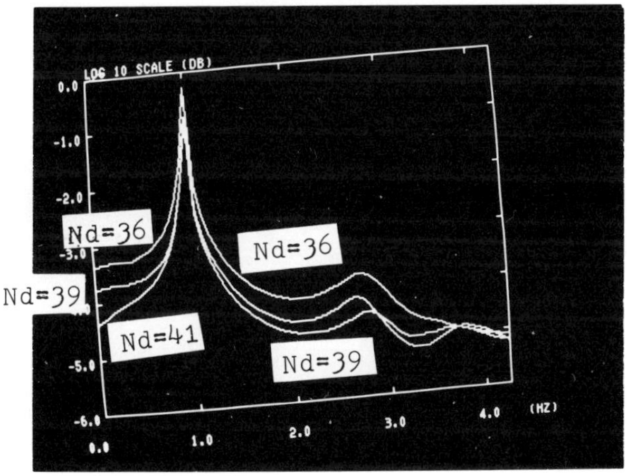

b) Same as Fig. 11a but with complex signals

Figure 11

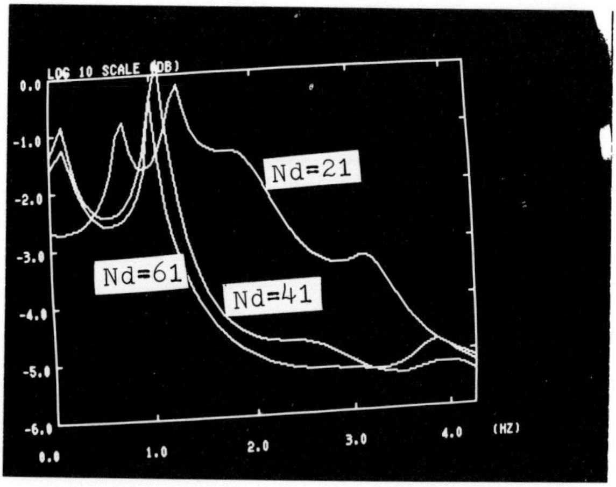

a) Maximum entropy power spectra of 2-channel real data (ϕ_1=45,ϕ_2=135, A=0.05, NPEF=9, N_d=21,41,61)

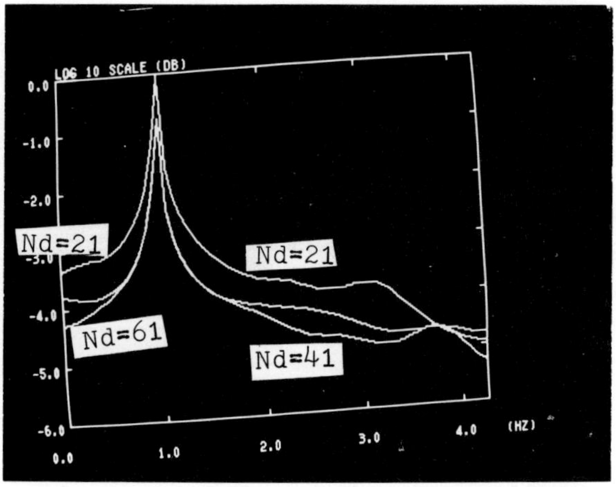

b) Same as Fig. 12a but with complex signals

Figure 12

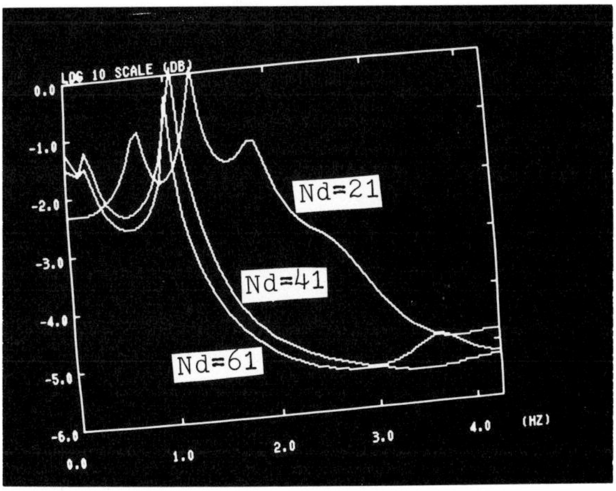

a) Maximum entropy power spectra of 2-channel real data (ϕ_1=-45, ϕ_2=-135, A=0.05, NPEF=9, N_d=21,41,61)

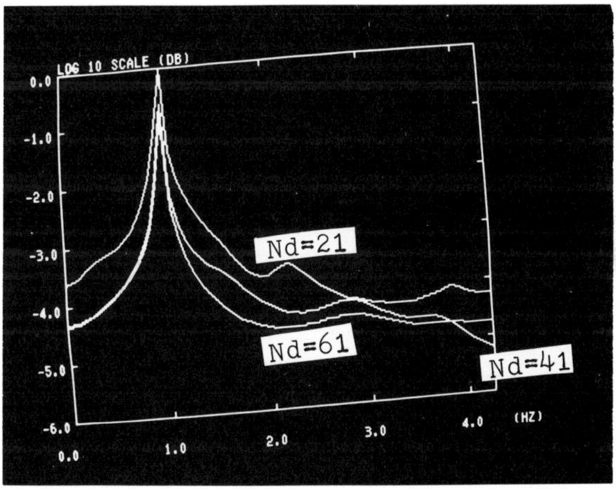

b) Same as Fig. 13a but with complex signals

Figure 13

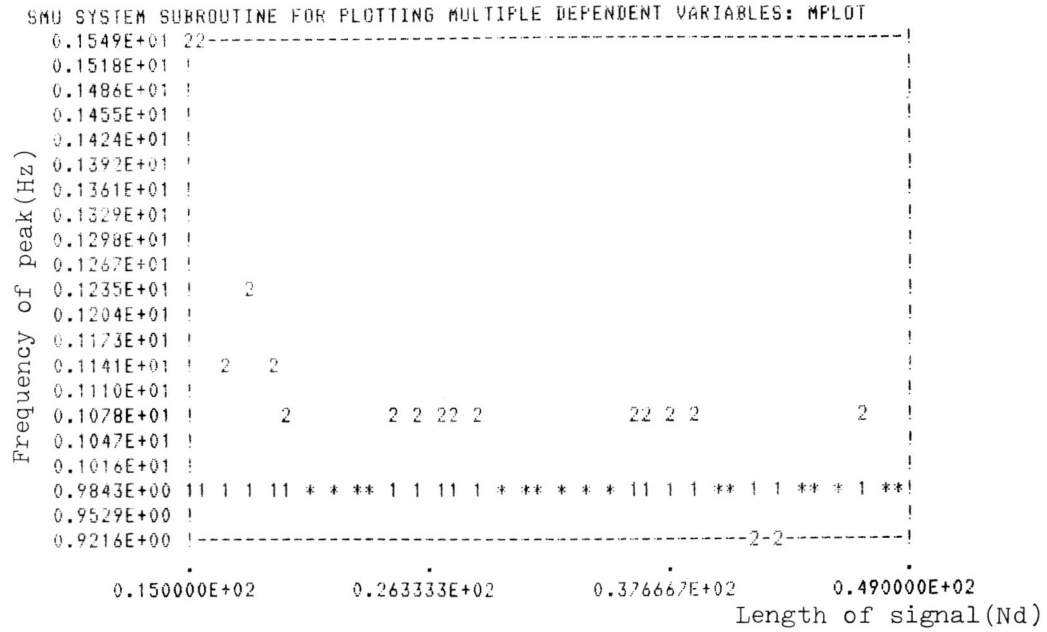

Figure 14 The location of spectral peaks for various lengths of 1Hz sinusoids. "1" denotes complex signals, "2" denotes real signals and "*" denotes both "1" and "2". $\phi_1=0$, $\phi_2=180$, A=0.05, NPEF=9.

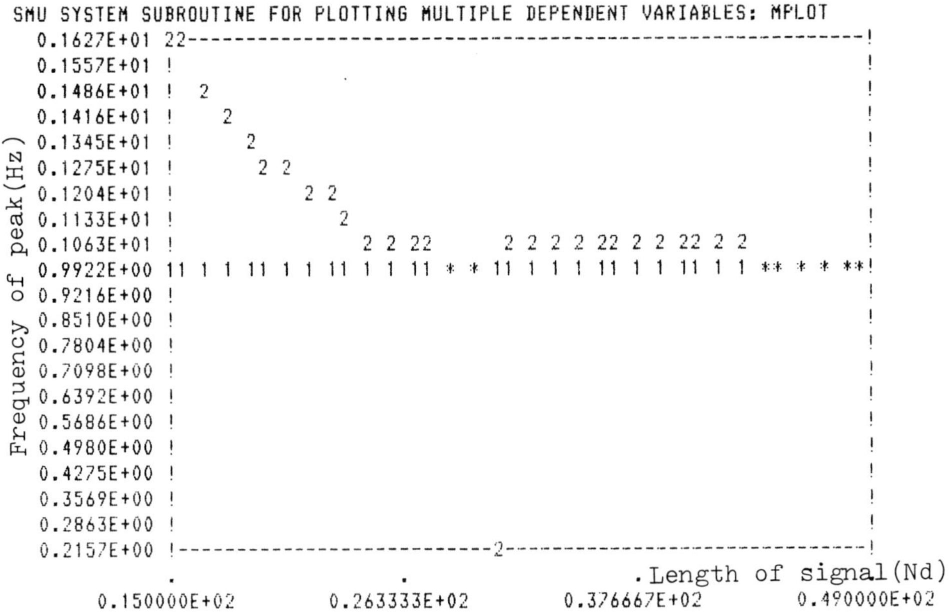

Figure 15 Same as Fig. 14 but with ϕ_1=45,ϕ_2=135,A=0.05, NPEF=9.

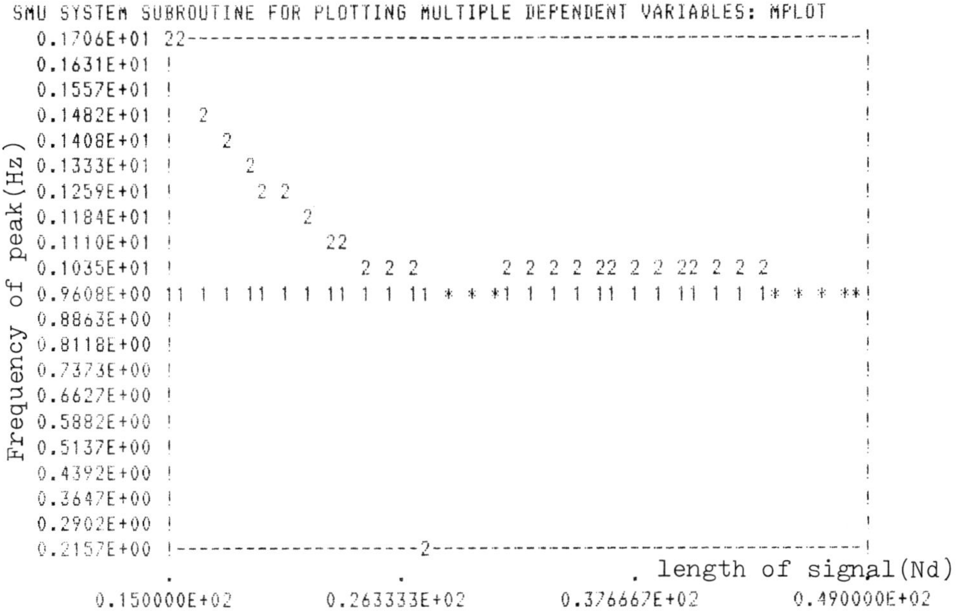

Figure 16 Same as Fig. 14 but with ϕ_1=-45,ϕ_2=-135,A=0.05, NPEF=9.

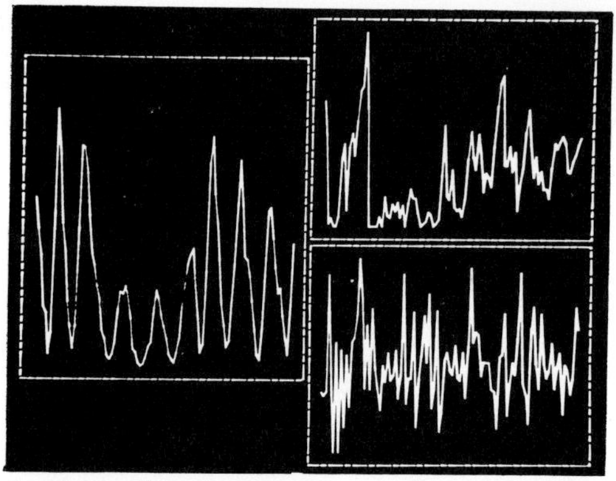

a) Plot of a 3-channel geophysical data

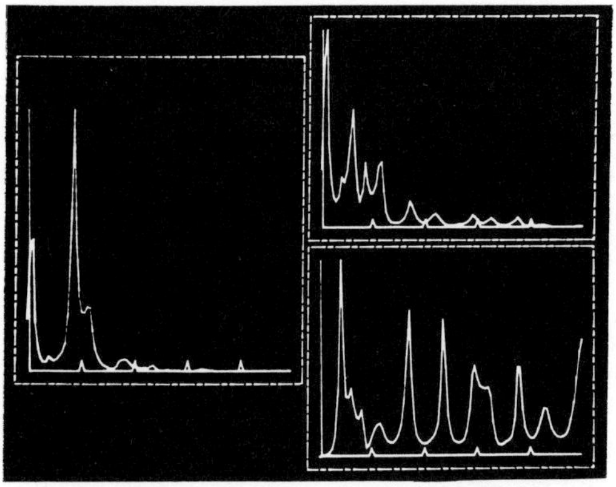

b) Maximum entropy power spectra

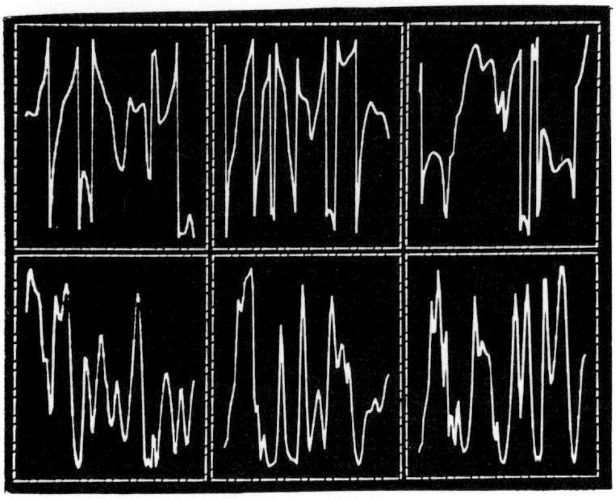

c) Between-channel phase differences

Figure 17

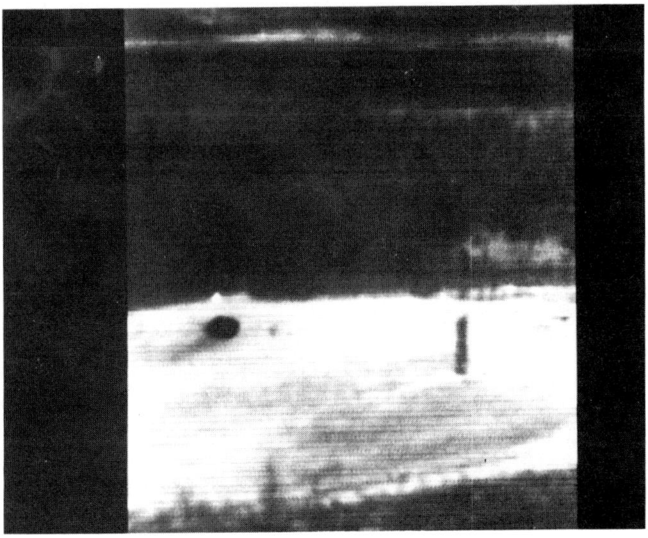

a) An infrared image

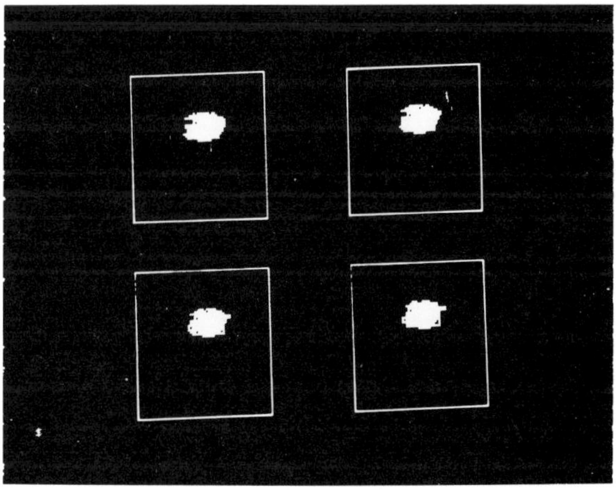

b) Segmentation of subregion by using multivariate maximum entropy spectral analysis

Figure 18

shows the phase differences between the two channels. The short version of the program (Appendix E) is used here. The number of lags is 17 which is determined as the optimum order. It is noted that the spectral peak for the sunspot numbers is determined accurately.

6. Image Segmentation Via Multivariate Spectral Analysis

The use of the multivariate autoregressive analysis for texture classification and segmentation was considered by Gambotto and Gueguen[12]. Their segmentation result, however, leaves much to be desired. We have employed the multivariate autoregression estimation program of Jones [6] on the infrared image shown in Fig. 18a. A small area of 64x64 is considered in the computer study. The subimage is processed by small squares of 2x10 pixels each, making use of the two-channel multivariate autoregression analysis. The R matrix is computed from learning samples. Then the mean-square prediction error defined as the trace of the matrix P_N given by Eq. (9) is computed and compared with a threshold. Each pixel is classified as belonging to an object or background. The idea is that the prediction error tends to be large for pixels near the object boundary and it is small for homogeneous regions. Segmentation is performed by pixel classification using the prediction error. The classified result superimposed on the binary image of original picture is shown in Fig. 18b for orders 4,5, 6 and 7. These results are acceptable but not as good as compared with the segmentation using Fisher's linear discriminant [13]. Furthermore, the amount of computation required for the multivariate autoregression analysis is much larger. Further study of image analysis using multivariate autoregression is much needed.

REFERENCES

1. Whittle, P., "On the fitting of multivariate autoregressions, and the approximate canonical factorization of a spectral density matrix," Biometrika, vol. 50, pp. 129-134, 1963.
2. Robinson, E. A., "Mathematical development of discrete filters for the detection of nuclear explosions," Journal of Geophysical Research, vol. 68, pp.5559-5567, 1963.
3. Wiggins, R. A., and E. A. Robinson, "Recursive solution to the multichannel filtering problem," Journal of Geophysical Research, vol. 70, pp.1885-1891, 1965.
4. Burg, J. P., "Maximum entropy spectral analysis," Ph.D. thesis, Dept. of Geophysics, Stanford University, 1975.
5. Morf, M., A. Vieira, D.T.L. Lee and T. Kailath, "Recursive multichannel maximum entropy spectral estimation," IEEE Trans. on Geoscience Electronics, vol GE-16, no.2., pp. 85-94, April 1978.
6. Jones, R. H., "Multivariate autoregression estimation using residuals," in "Applied Time Series Analysis," edited by D. F. Findley, Academic Press, 1978.
7. Nuttall, A. H., "Fortran program for multivariate linear predictive spectral analysis, employing forward and backward averaging," NUSC Technical Document 5419, Naval Underwater Systems Center, New London, Connecticut 1976.

8. Strand, O. N., "Multichannel complex maximum entropy (autoregressive) spectral analysis," IEEE Trans. on Automatic Control, vol. AC-22, no. 4, pp. 634-640, August 1977.

9. Fougere, P. F., "Spontaneous line splitting in multichannel maximum entropy power spectra," Proceedings of the First ASSP Workshop on Spectral Estimation, Hamilton, Ontario, Canada, PP. 161-163, Aug. 17, 18, 1981.

10. Strand, O. N., "Computer programs for maximum entropy spectral analysis of real and complex multichannel time series (with microfilm plots)," NOAA Technical Memorandum ERL WPL-22, Boulder, Colorado April 1977.

11. Robinson, E. A., "Multichannel Time Series Analysis with Digital Computer Programs," Holden-Day, 1967.

12. Gambotto, J. P. and C. Gueguen, "A multidimensional modelling approach to texture classification and segmentation", Proc. of the IEEE International Conference on Acoustics, Speech and Signal Processing, Washington, D. C., pp. 962-966, April 1979.

13. Chen, C. H. and C. Yen, "Object isolation in FLIR images using Fisher's linear discriminant", Pattern Recognition Journal, vol. 15, May 1982.

CHAPTER 6

On a Two-Dimensional Maximum Entropy Spectral Analysis Method with Application to Texture Analysis

1. Introduction

Although it is generally recognized that texture images contain statistical, spectral and structural domain information, the use of spectral information alone can be quite effective in texture-image analysis studies such as texture discrimination and segmentation. Bajcsy and Liberman [1] express the power spectrum in polar coordinates, then integrate over r and ϕ to obtain the two one-dimensional functions. The location of peaks in these functions indicates prominant texture coarseness and directionality. Weszka et. al. [2] integrated the power spectrum within 16 spatial frequency zones which were combinations of four 1-octave frequency ranges and four 45° orientation sectors. They also computed eight "contrast" measures based on the co-occurrence matrix, and obtained better discrimination than with the power spectrum measures. Laws [3] computed a number of energy measures by filtering the texture with sets of small linear operators, then squaring and summing the output of each filter. He reported better discrimination with the energy than with the co-occurrence measures.

A fundamental problem with the two-dimensional power spectrum analysis is the computational accuracy and computational complexity. For a texture study, accurate power spectrum must be computed from the small image segments. In this case, the two-dimensional Fourier analysis cannot provide sufficient accuracy as the Fourier analysis is accurate only with a large number of pixels. The two-dimensional maximum entropy spectral analysis, however, is very suitable for a small number of pixels. The computational complexity has been a drawback in using the two-dimensional maximum entropy spectral estimation procedures. Recently, Lim and Malik [4, 5, 6] have proposed an efficient iterative algorithm for the two-dimensional maximum entropy power spectrum estimation suitable for a minicomputer implementation. Their method has been adapted and generalized for use on our PDP 11/45 minicomputer for the texture-image analysis. A three-dimensional graphics software is developed for the spectral display at different viewing positions. Extensive computer results on the spectral analysis of texture images are reported in Section 3.

2. Two-Dimensional Power Spectrum Estimation

To obtain the power spectrum of a two-dimensional signal, the direct method is to calculate the two-dimensional Fourier transform of the autocorrelation function, i.e.:

$$P_x(\omega_1, \omega_2) = \sum_{n_1=-\infty}^{\infty} \sum_{n_2=-\infty}^{\infty} R_x(n_1, n_2) e^{-j(n_1\omega_1 + n_2\omega_2)} \tag{1}$$

The following notations will be used in this chapter.

$x(n_1, n_2)$: A 2-D random signal whose power spectrum we wish to estimate.

$R_x(n_1, n_2)$: Autocorrelation function of $x(n_1, n_2)$

$\hat{R}_x(n_1, n_2)$: An estimate of $R_x(n_1, n_2)$

$P_x(\omega_1, \omega_2)$: Power spectrum of $x(n_1, n_2)$

$\lambda(n_1, n_2)$: Autocorrelation function whose power spectrum is $1/P_x(\omega_1, \omega_2)$

A : A set of points (n_1, n_2) for which $R_x(n_1, n_2)$ is known.

F : Discrete Fourier Transform

F^{-1} : Inverse Discrete Fourier Transform

From Eq. (1), it is important to note that the determination of the power spectral density entails complete knowledge of the autocorrelation function of generally infinite extent. For the finite signal, it has been shown that the method has poor resolution, due to the truncated sampled autocorrelation function set. For the one-dimensional case, the power spectrum expressed by Eq. (3) of Chapter 2 can be determined by using Levinson recursion and Burg's algorithm to solve for the filter coefficients in the normal equations. This is the standard procedure for the Burg's maximum entropy spectral analysis. But for the two-dimensional case, the problem is different since the normal equations become a highly nonlinear problem [4]. For the general form of the two-dimensional case,

$$\sum_{(i,j)\varepsilon B}\sum a_{ij} R_x(r-i, s-j) = R_x(r, s) \qquad \text{for } (r, s) \ \varepsilon \ B \tag{5}$$

Here the set B consists of all points where the filter mask has nonzero values, and the power spectrum obtained from a_{ij} is given by

$$\hat{P}_x(\omega_1, \omega_2) = \frac{1}{\left| \sum_{(k,l)\varepsilon B}\sum a_{kl} \exp(-j\omega_1 k - j\omega_2 l) \right|^2} \tag{6}$$

In this case, we can see from Eq. (5) that the size of independent values of $R_x(n_1, n_2)$ required to solve the above set of equations is greater than the size of the filter mask. For example, Fig. 1(a) shows an autoregressive filter mask size of 3x2, and Fig. 1(b) shows a larger size of independent values of $R_x(n_1, n_2)$ required to solve for a_{ij} in Fig. 1(a) by Equation (5).

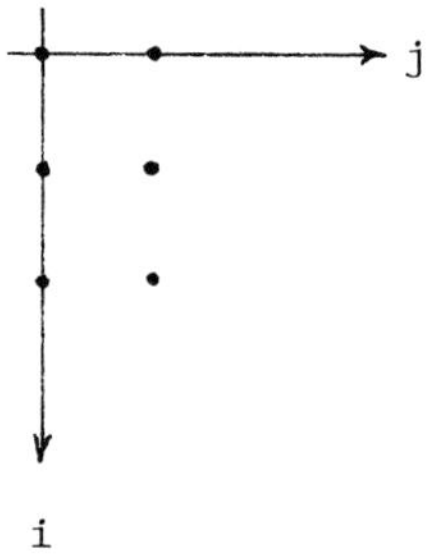

Fig. 1a

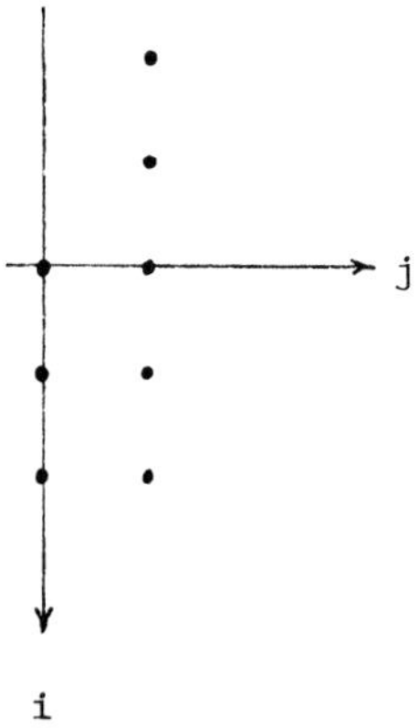

Fig. 1b

Clearly, the number of correlation points needed is greater than the number of filter coefficients. Since the estimated power spectrum given by Eq. (6) is completely determined by the coefficients alone, it does not possess enough degrees of freedom to solve for the spectrum. Therefore, the normal equations are not linear as in the one-dimensional case. Many methods have been proposed to extend the Levinson's and the Burg's algorithms to two dimensions. However, those algorithms are not computationally attractive, and there is no guarantee that a solution or an approximate solution can be obtained. For instance, Burg [7] has proposed an iterative solution which requires the inversion of a matrix in each iteration where the dimension of the matrix is of the order of the number of the given autocorrelation points. No experimental results using this technique have yet been reported. Wernecke and D'Addario [8] have proposed a scheme in which an attempt is made numerically to maximize the entropy. The maximization is done by continuously adjusting the power spectrum estimate and evaluating the expressions for the entropy and its gradient. The procedure is computationally expensive and is not guaranteed to have a solution. Woods [9] expresses the maximum entropy method as a power series in the frequency domain and attempts to approximate the maximum entropy spectrum estimate by truncating the power series expansion. Even though such an approach has some computational advantages relative to others, the method is restricted to the class of signals for which the power series expansion is possible.

As a closed form solution of the two-dimensional maximum entropy method is hard to obtain, Lim and Malik developed a new iterative algorithm, using adaptive filtering concept. This algorithm can correctly estimate the true spectrum and is computationally simple due to the utilization of Fast Fourier Transform (FFT). The basic idea of this algorithm is derived from the notion that the given correlation points in region A must be consistent and the corresponding coefficient should be zero outside region A. Here A is the set of points where the autocorrelation is known. This iterative process is repeated until the optimal solution is obtained. More specifically, given $R_x(n_1, n_2)$ for $(n_1, n_2) \varepsilon A$, we determine $\hat{P}_x(\omega_1, \omega_2)$ as

$$\hat{P}_x(\omega_1, \omega_2) = \frac{1}{\sum\sum_{(n_1,n_2)\,\varepsilon\, A} \lambda(n_1,n_2) e^{-j\omega_1 n_1} e^{-j\omega_2 n_2}} \tag{7}$$

and

$$R_x(n_1, n_2) = F^{-1}[\, P_x(\omega_1, \omega_2)\,] \quad \text{for } (n_1, n_2)\,\varepsilon\, A.$$

A simple flowchart is shown in Fig. 2. We begin with some initial estimate of $\lambda(n_1, n_2)$, obtain the corresponding correlation function, correct the resulting correlation function for $(n_1, n_2) \varepsilon A$ with the known $R_x(n_1, n_2)$, obtain the corresponding $\lambda(n_1, n_2)$ from the correct correlation function, and then replace the resulting $\lambda(n_1, n_2)$ with 0 for $(n_1, n_2) \notin A$. This completes one iteration and the corrected $\lambda(n_1, n_2)$ is a new estimate of $\lambda(n_1, n_2)$.

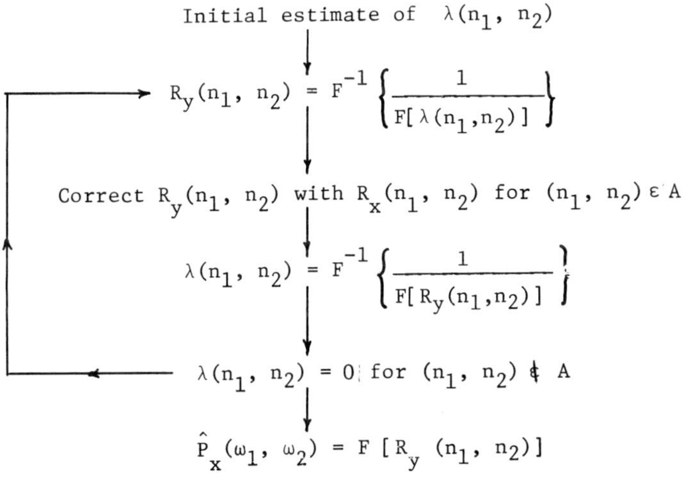

Fig. 2

However, to prevent the zero crossing problem when taking the inverse of $F[\lambda(n_1, n_2)]$ and $F[R_y(n_1, n_2)]$ due to the correction of $R_y(n_1, n_2)$ and $\lambda(n_1, n_2)$, and also to keep the correlation function positive definite, Lim and Malik modified the procedure by linearly interpolating some parameters to prevent the divergence of the error. An increased rate of convergence was also aimed at ([4], Fig. 4). Thus some constraints have been incorporated to get a practical algorithm.

3. Examples

The following are some examples of using Lim and Malik's algorithm which has been implemented in our PDP 11/45 minicomputer. The computer program is given in Appendix F. The first input signal is a two-dimensional autocorrelation function originated from sinusoids buried in white noise, so that the correlation function given has the form of

$$R_x(n_1 , n_2) = \sigma^2 \delta(n_1, n_2) + \sum_{i=1}^{M} a_i^2 \cos(\omega_{i1} n_1 + \omega_{i2} n_2) \qquad (8)$$

where σ^2 is the white noise power, M is the number of sinusoids, $a_i^2/2$ is the power of the ith sinusoid, and ω_{i1} and ω_{i2} represent the frequency of the ith sinusoid. Fig. 3 - Fig. 5 show the spectrum of the signals with 1, 2, 3 sinusoids respectively, which correspond to the input data shown in Tables 1, 2, 3. From these results, we can see that this iterative procedure can easily predict the true spectrum although the matrix of the autocorrelation function is not very large.

However, in Lim and Malik's examples and the examples shown in Figs. 3-5, the expression used for the autocorrelation function calculation is Eq. (8) which has an analytical form. This expression is based on the consideration of continuous and infinite sinusoids. So its values are completely symmetric. For the practical two-dimensional signal, this is no longer the case. To generalize this algorithm so that it can be used for any two-dimensional signal, we use the well-known unbiased formula for the calculation of the autocorrelation of real image data, i.e.,

$$R(n_1, n_2) = \frac{1}{NxN} \sum_{k=1}^{N-n_1} \sum_{j=1}^{N-n_2} x(k,j)\, x(k+n_1, j+n_2) \qquad (9)$$

To make sure that this expression can fit the real autocorrelation function, we have used it to generate a simulated autocorrelation function of two-dimensional sinusoids, and the estimated spectrum is almost the same as that calculated from Eq. (8).

In this section, we will present the results of the two-dimensional maximum entropy power spectrum estimation of the real texture images. This data is taken from the U. S. C. data base. The size of each image is 64x64. The original textures are shown in Fig. 6 . These pictures will reappear but 5 times larger in Fig. 7a - Fig. 15a with their corresponding estimated spectra shown in Fig. 7b - Fig. 15b, Fig. 7c - Fig. 15c and Fig. 7d - Fig. 15d.

Fig. 7b shows that the picture has one main frequency component near (0.01, 0.22), one small dc component and one frequency at (0.0, 0.5) and other very small ripples. Because of the three-dimensional display, Fig. 7c and Fig. 7d cannot accurately indicate the (0.01, 0.22) point, but we can see the entire power spectrum function distribution in Fig. 7c and Fig. 7d. Fig. 8b shows that the main frequency component is approximately equal to (0.02, 0.126) and other small ripples

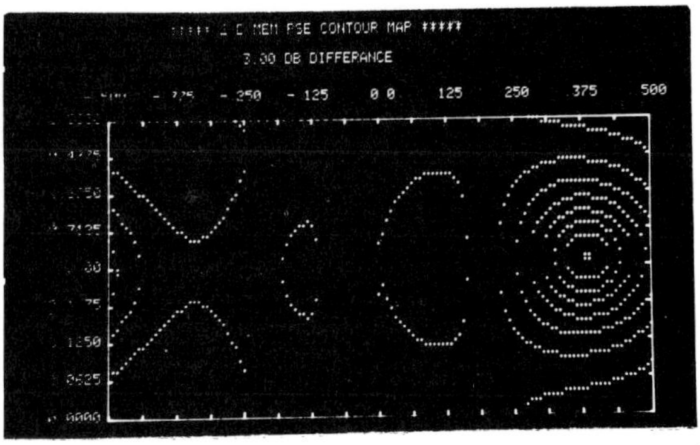

a) The contour map of the estimated power spectrum with ΔdB=3. Note: ΔdB is the dB value difference between adjacent contours

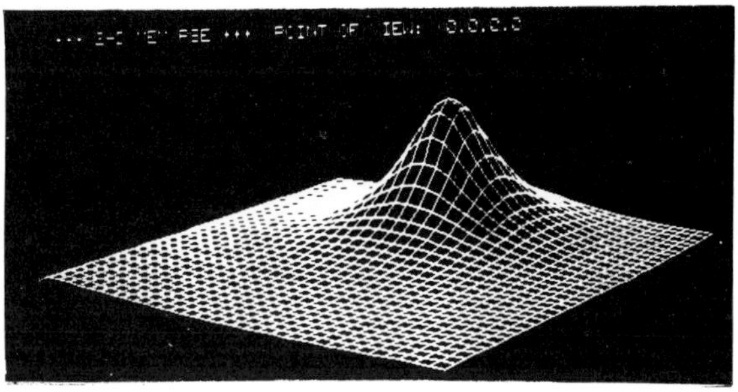

b) 3-D display of the estimated power spectrum. The view is from (ω_{i1}=0,ω_{i2}=0) point

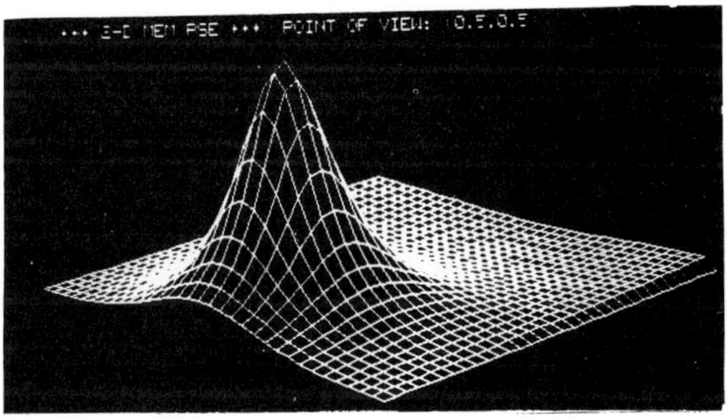

c) 3-D display of the estimated power spectrum as viewed from (ω_{i1}=0.5,ω_{i2}=0.5) point

Figure 3

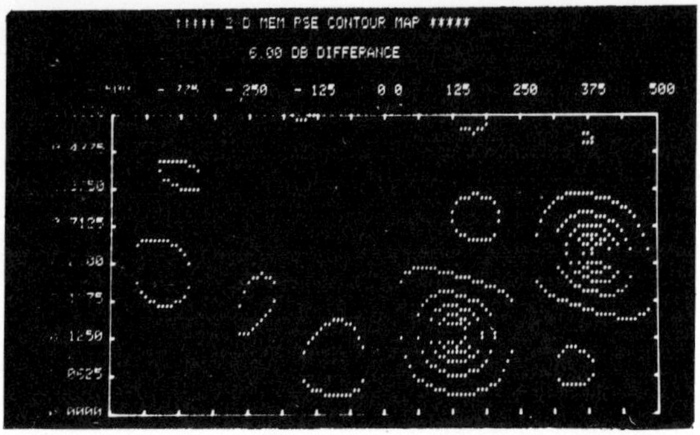

a) The contour map of the estimated power spectrum with ΔdB=6

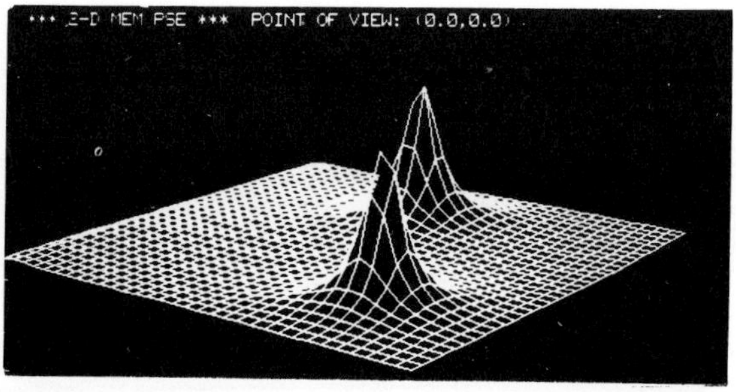

b) 3-D display of the estimated spectrum at (0.0,0.0) point

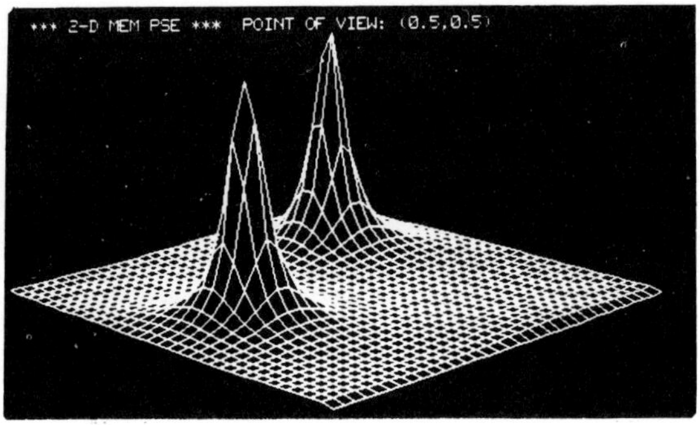

c) 3-D display of the estimated spectrum at (0.5,0.5) point

Figure 4

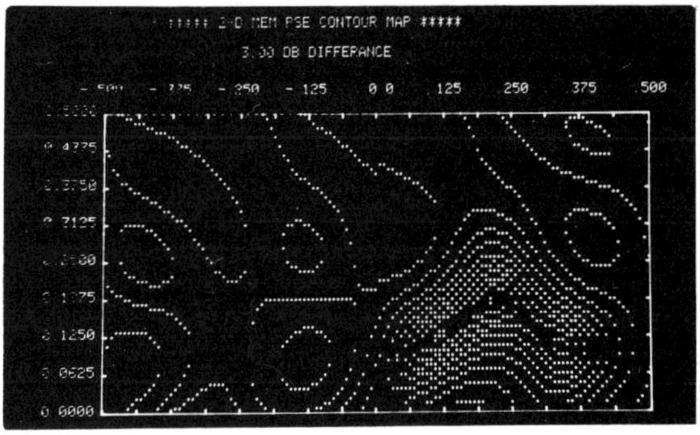

a) The contour map of the estimated power spectrum with ΔdB=3

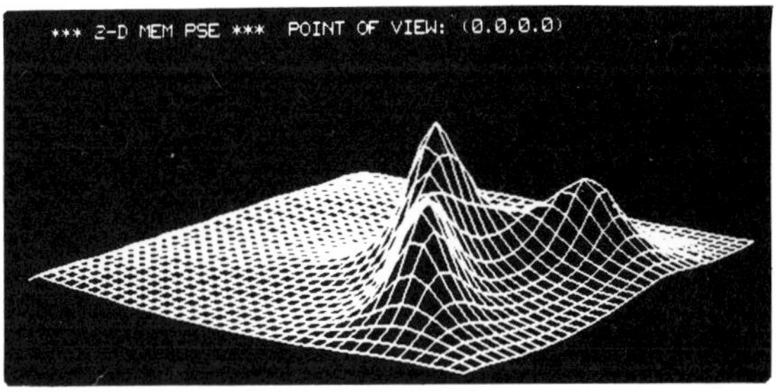

b) 3-D display of the estimated spectrum at (0.0,0.0) point

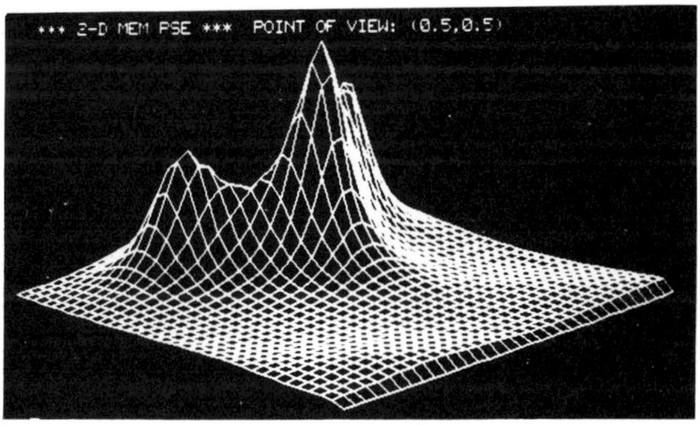

c) 3-D display of the estimated power spectrum at (0.5,0.5) point

Figure 5

A	M	σ^2	a_i^2	$(\omega_{i1}/2\pi,\ \omega_{i2}/2\pi)$	ε	NDFT	NITR
5x5	1	5.0	1.0	0.375,0.250	10^{-4}	32	5

TABLE I

A	M	σ^2	a_i^2	$(\omega_{i1}/2\pi,\ \omega_{i2}/2\pi)$	ε	NDFT	NITR
7x7	2	6.0	1.0	0.125,0.125	10^{-3}	32	12
7x7	2	6.0	1.0	0.375,0.250	10^{-3}	32	12

TABLE II

A	M	σ^2	a_i^2	$(\omega_{i1}/2\pi,\ \omega_{i2}/2\pi)$	ε	NDFT	NITR
7x7	3	6.0	1.0	0.10,0.10	10^{-4}	32	21
7x7	3	6.0	1.0	0.30,0.10	10^{-4}	32	21
7x7	3	6.0	1.0	0.20,0.20	10^{-4}	32	21

TABLE III

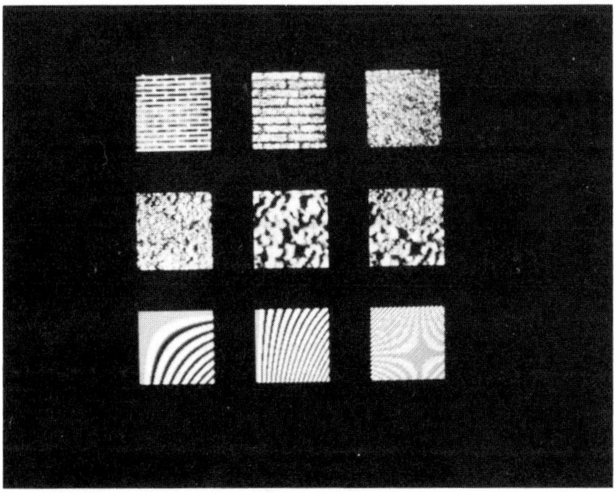

Figure 6 The original test texture data taken from the USC data base. Each data format is 64x64. The right one in second row is reconstructed from the left and center pictures of the second row.

a) The original data

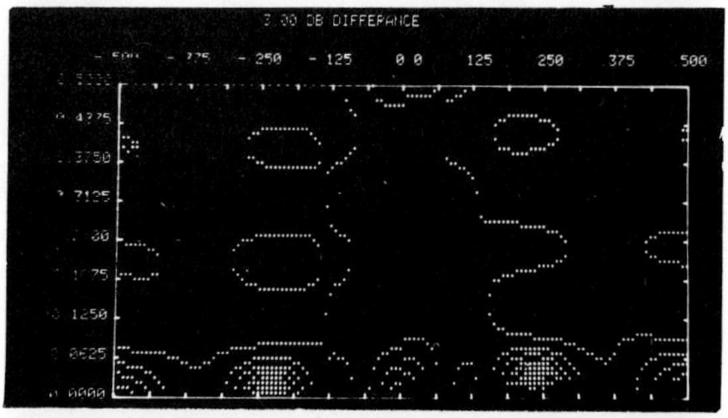

b) Contour map with ΔdB=3 and main frequency around (0.01,0.22)

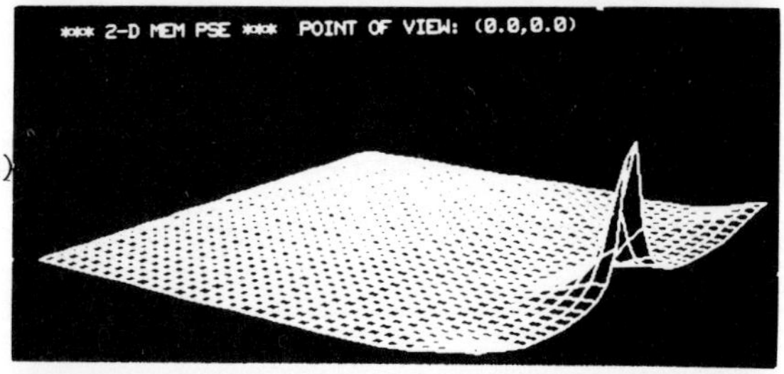

c) 3-D display at (0.0,0.0) point of view

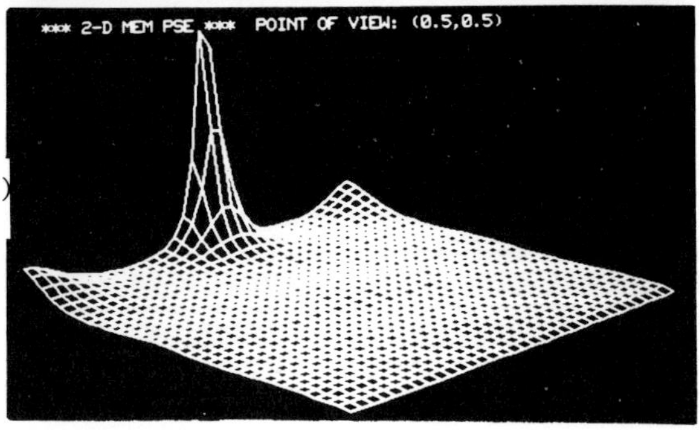

d) 3-D display at (0.5,0.5) point of view

Figure 7

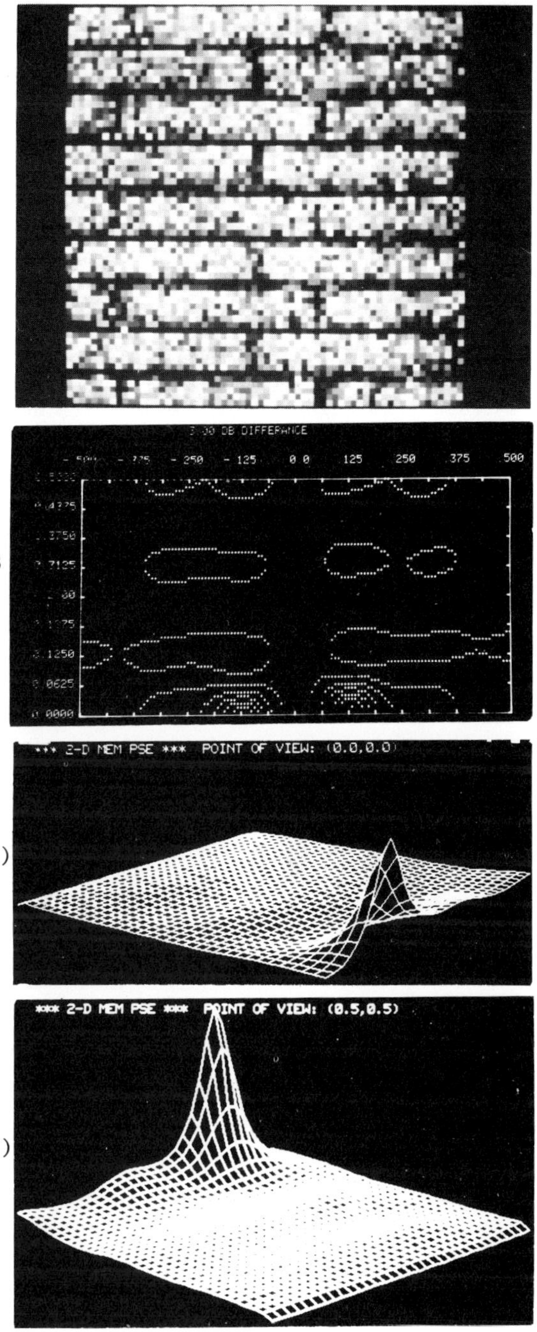

a) The original data

b) Contour map with ΔdB=3 and main frequency around (0.02,0.126)

c) 3-D display at (0.0,0.0) point of view

d) 3-D display at (0.5,0.5) point of view

Figure 8

are evident. Comparing Figs. 7 and 8, we can see clearly the difference in power spectrum for different textures.

Fig. 9 shows that the texture contains a main frequency component around (0.0, 0.0) and other small amplitude components. From the corresponding picture (upper right segment) in Fig. 6, we can see that it contains mostly high gray level pixels, which means that it has a high dc component. By taking a longer DFT length, we can predict the other small amplitude frequency terms. This effect can be seen in Fig. 12 which is a reconstructed picture using parts of Fig. 10 and Fig. 11. From Fig. 12d, we can see the estimated spectrum is the composition of Fig. 10d and Fig. 11d. The frequency term near (0.06, 0.125) is affected by the contribution of Fig. 11. But when we take a larger autocorrelation function matrix and longer DFT length, this term can be recovered. Fig. 16 clearly demonstrates this phenomenon. To prove the accuracy of the two-dimensional maximum entropy power spectrum estimation algorithm, three sets of "nearly periodic" texture data were tested. From Figs. 13-15, the results obtained are satisfactory and appear to give the real power spectrum. Fig. 17 shows the spectrum of a 64x64 subimage of a Seasat image.

4. Discussion

For a real two-dimensional signal, this algorithm can accurately predict the main frequency components with a moderate size autocorrelation function (ACF) matrix and short DFT length. However, for the other frequency terms with smaller amplitude, it must take a larger ACF matrix and longer DFT length to discriminate, since the larger ACF matrix will add more information about the signal. But this would take too much computational time. Also, close spectral peaks cannot be resolved by this algorithm with a moderate size ACF matrix and DFT length. To understand and to improve the spectral resolution capability of the algorithm, the true spectrum must be analyzed. We are now continuing the research by generating an artificial two-dimensional signal which is predicted by the coefficients of the AR model suggested by Cadzow and Ogino [10]. It can be expressed as

$$x(n_1, n_2) = -\sum_{k=o}^{N}\sum_{m=o}^{N} a_{km}\, x(n_1-k, n_2-m)$$

where the real power spectrum can be calculated by using

$$P_x(\omega_1, \omega_2) = \frac{1}{\left|\sum_{k=o}^{N}\sum_{m=o}^{N} a_{km} e^{-j(k\omega_1+m\omega_2)}\right|^2} \qquad (11)$$

We expect that by comparing the estimated spectrum with the true one calculated from Eq. (11), we will be able to find some special properties which can be used as a reference for improvement.

For the different types of texture images studied in this monograph, the two-dimensional maximum entropy spectrum method has provided a reasonably good spectral discrimination. For the small image size considered, the maximum entropy spectrum is indeed far superior to

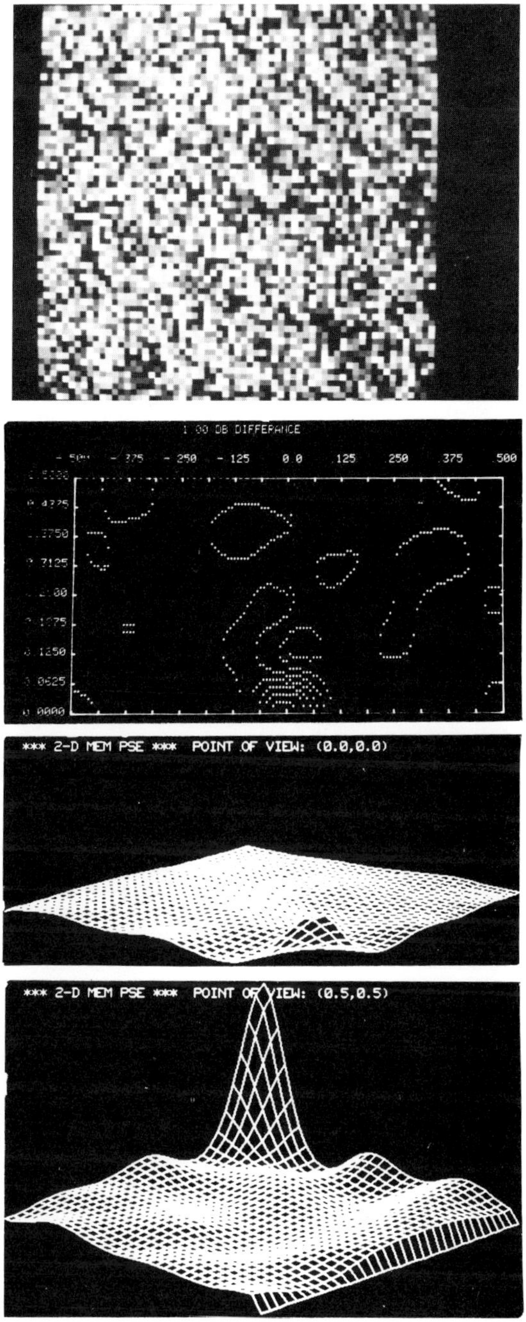

a) The original data

b) Contour map with ΔdB=1. and main frequency around (0.0,0.0)

c) 3-D display at (0.0,0.0) point of view

d) 3-D display at (0.5,0.5) point of view

Figure 9

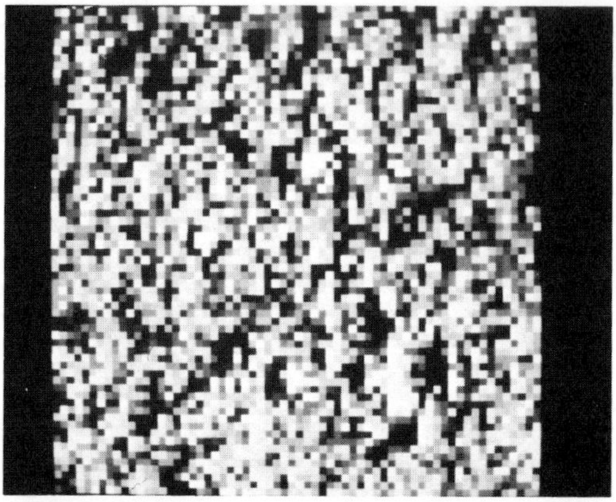

a) The original data

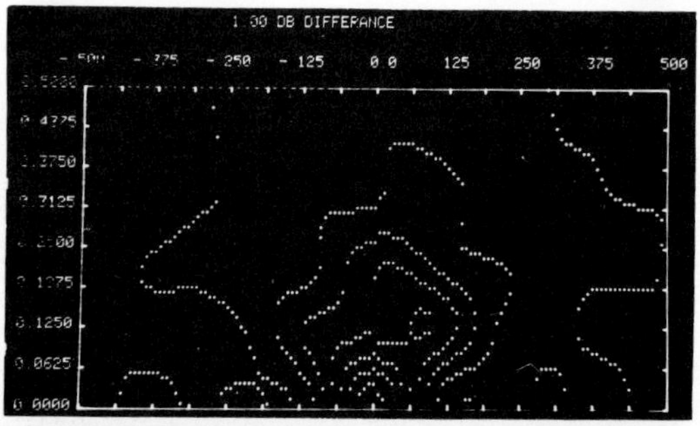

b) Contour map with ΔdB=1.0 and main frequencies around (0.0,0.0) and (0.06,0.125)

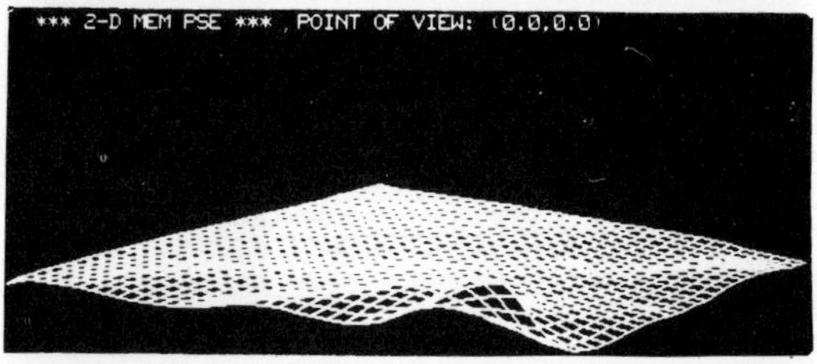

c) 3-D display at (0.0,0.0) point of view

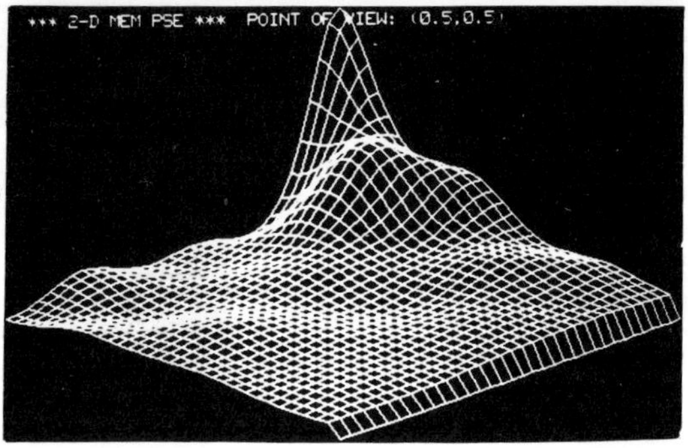

d) 3-D display at (0.5,0.5) point of view

Figure 10

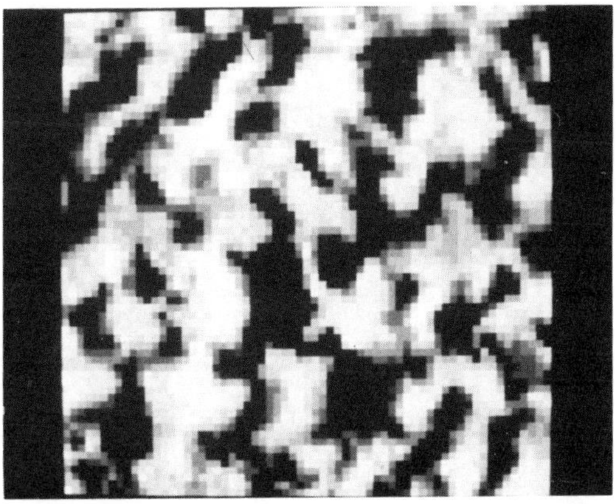

a) The original data

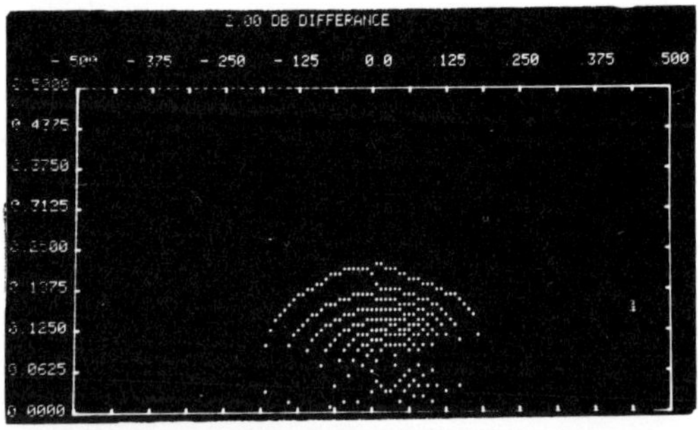

b) Contour map with ΔdB=2. and main frequency around (0.03,0.07)

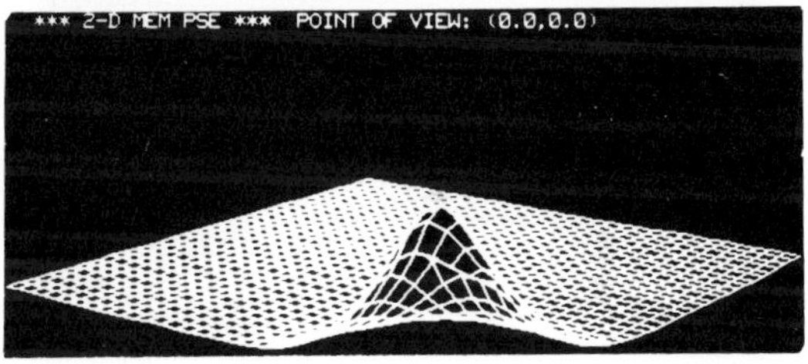

c) 3-D display at (0.0,0.0) point of view

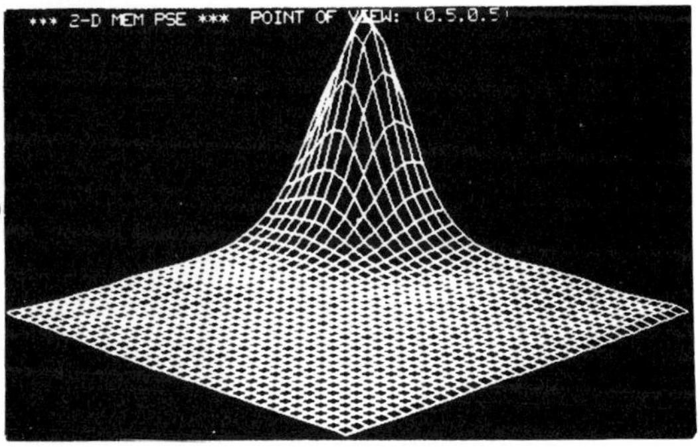

d) 3-D display at (0.5,0.5) point of view

Figure 11

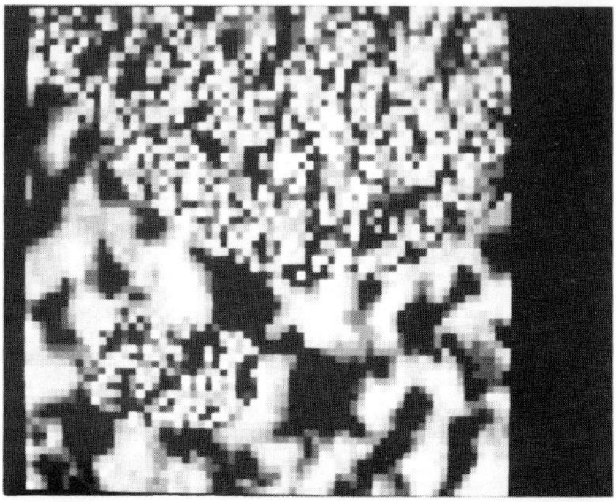

a) The data as constructed from Figs. 10a and 11a

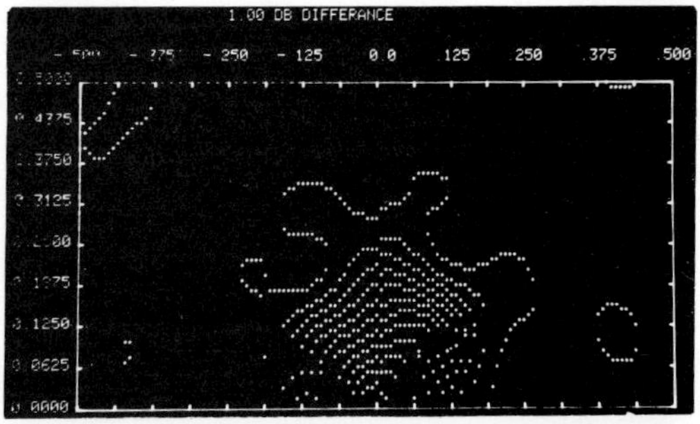

b) Contour map with ΔdB=1. and main frequency around (0.0,0.0)

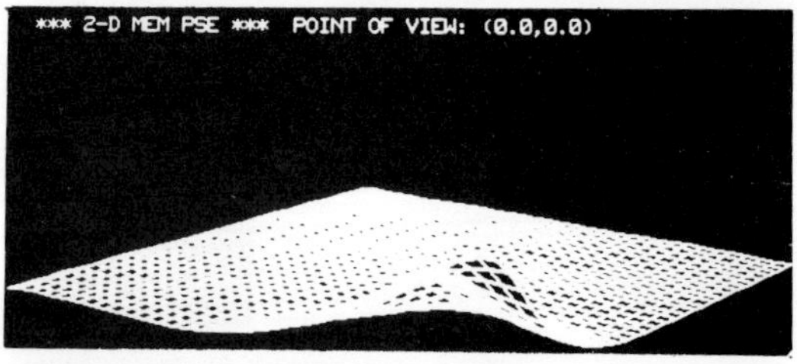

c) 3-D display at (0.0,0.0) point of view

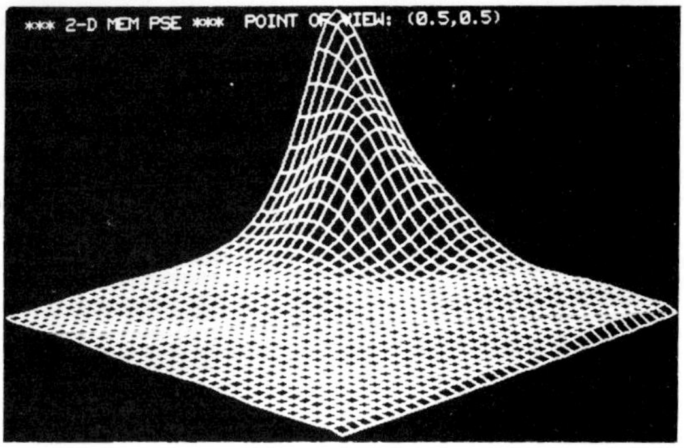

d) 3-D display at (0.5,0.5) point of view

Figure 12

a) The original data

b) Contour map with ΔdB=3.
and main frequency around
(0.08,0.09)

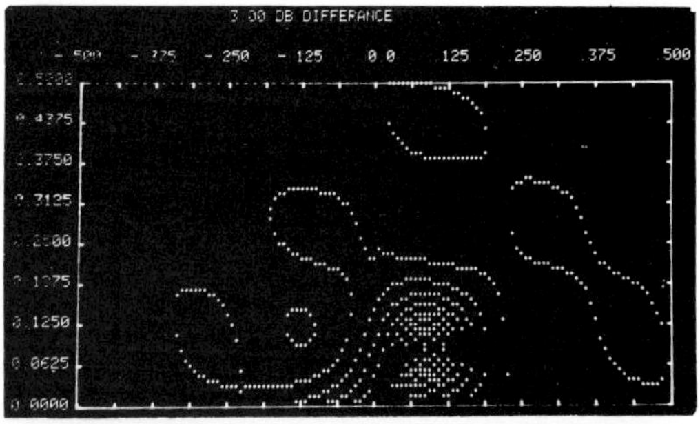

c) 3-D display at (0.0,0.0)
point of view

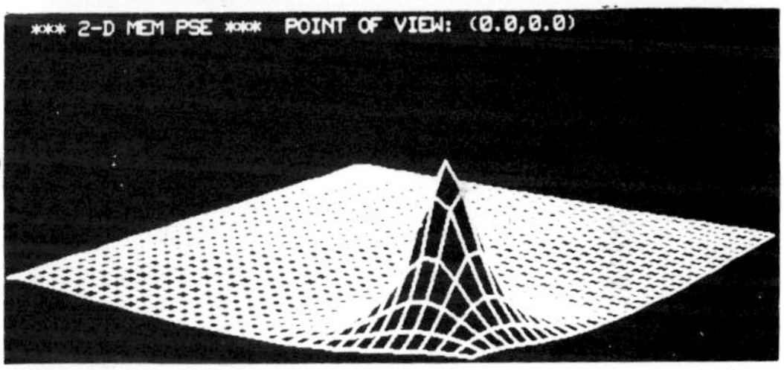

d) 3-D display at (0.5,0.5)
point of view

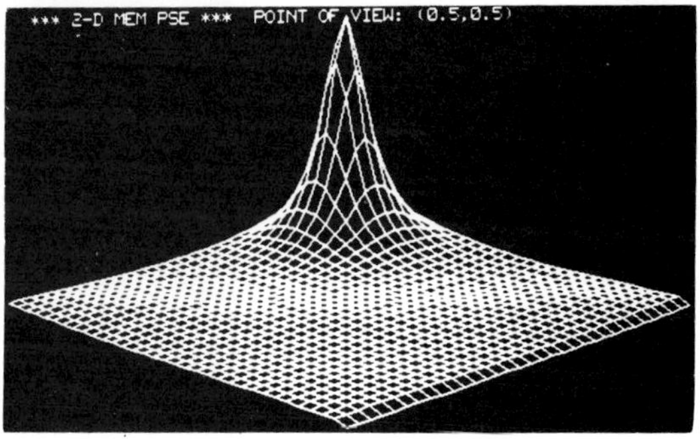

Figure 13

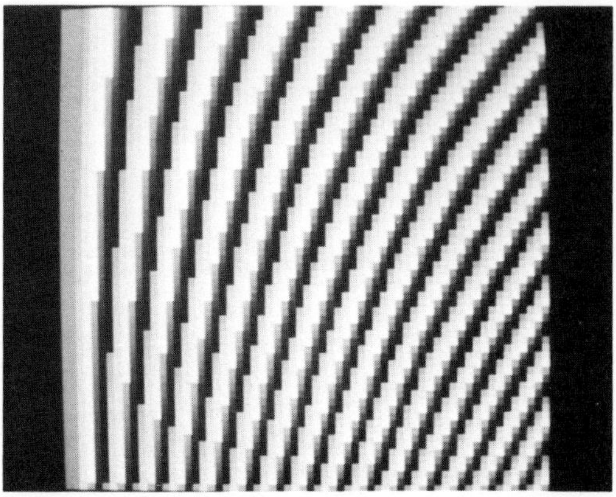

a) The original data

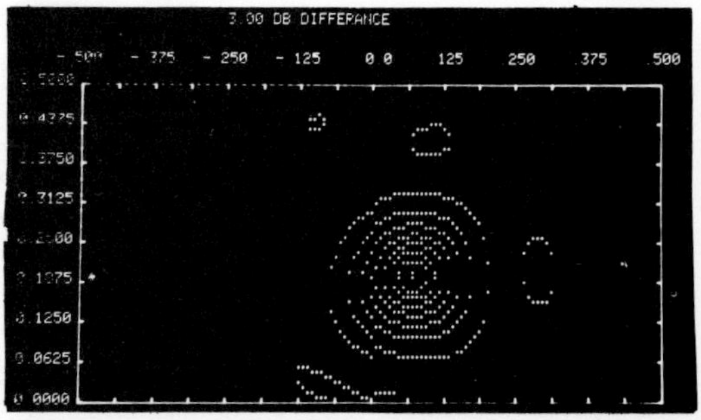

b) Contour map with ΔdB=3. and main frequency around (0.065,0.20)

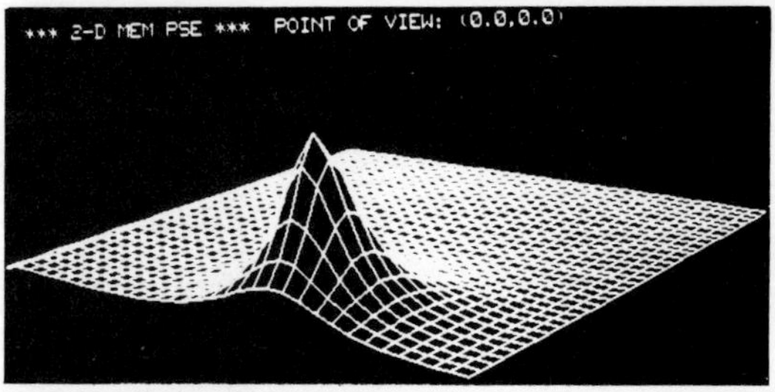

c) 3-D display at (0.0,0.0) point of view

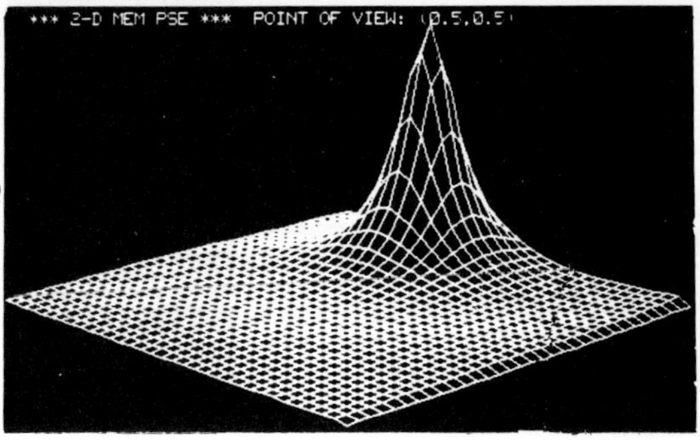

d) 3-D display at (0.5,0.5) point of view

Figure 14

a) The original data

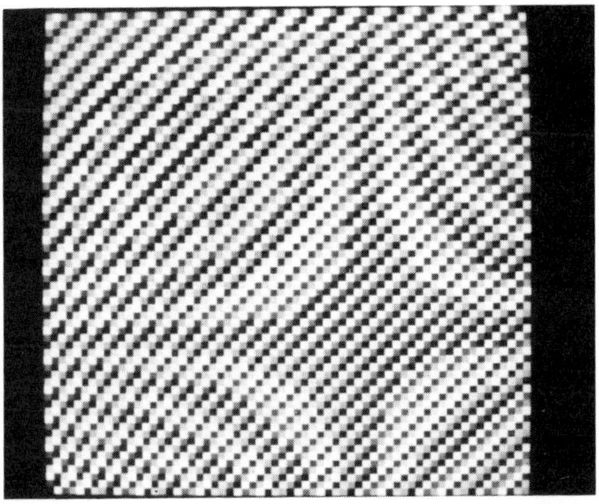

b) Contour map with ΔdB=2. and main frequency around (0.3125,0.3125)

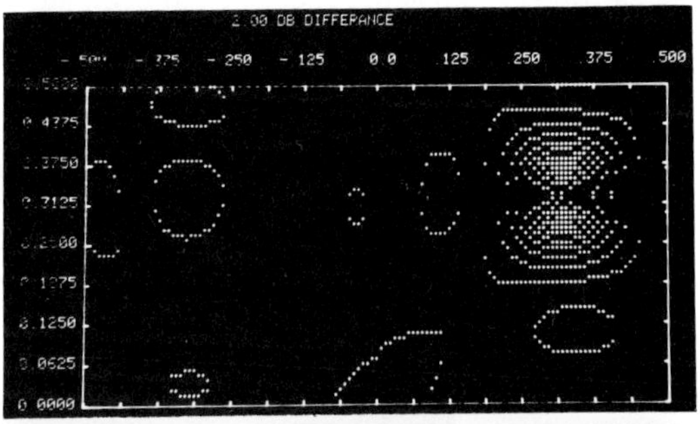

c) 3-D display at (0.0,0.0) point of view

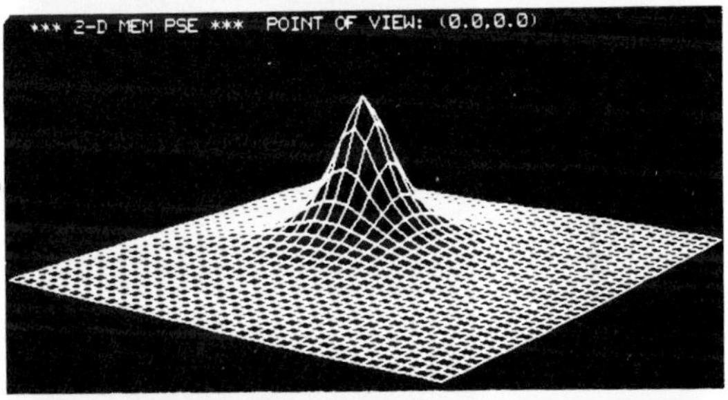

d) 3-D display at (0.5,0.5) point of view

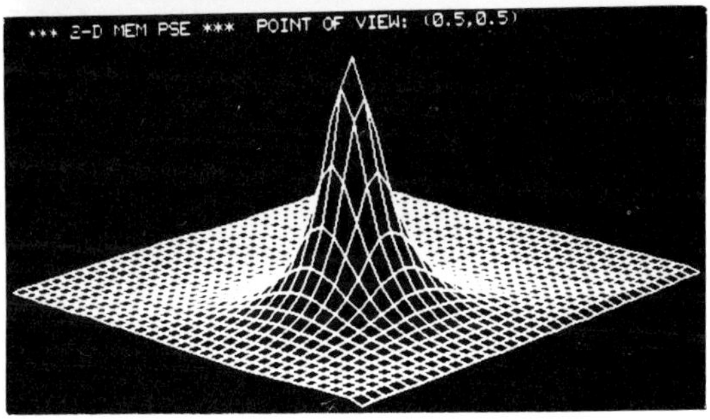

Figure 15

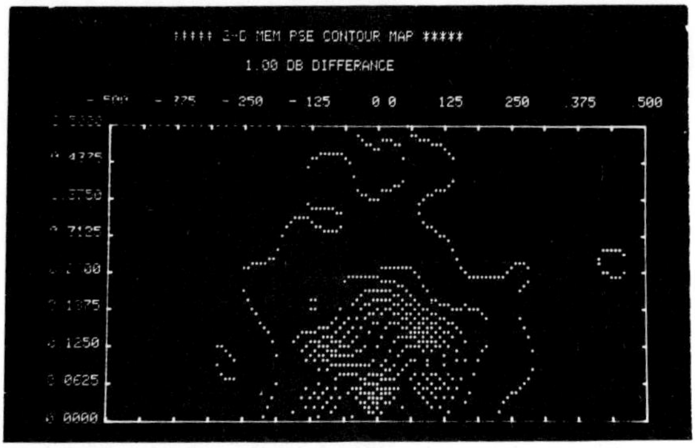

a) The contour map of the maximum entropy spectrum of Fig. 12a with a larger autocorrelation matrix. The main frequencies are around (0.01,0.03) and (0.07,0.125)

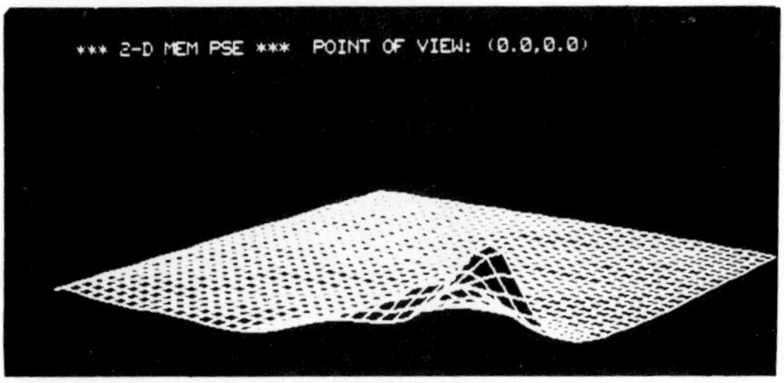

b) 3-D display at (0.0,0.0) point of view

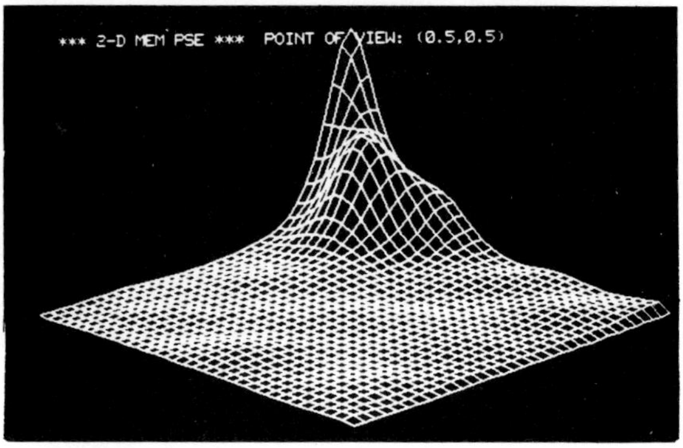

c) 3-D display at (0.5,0.5) point of view

Figure 16

a) A Seasat image and a 64x64 subimage considered (right) for 2-D maximum entropy spectral analysis

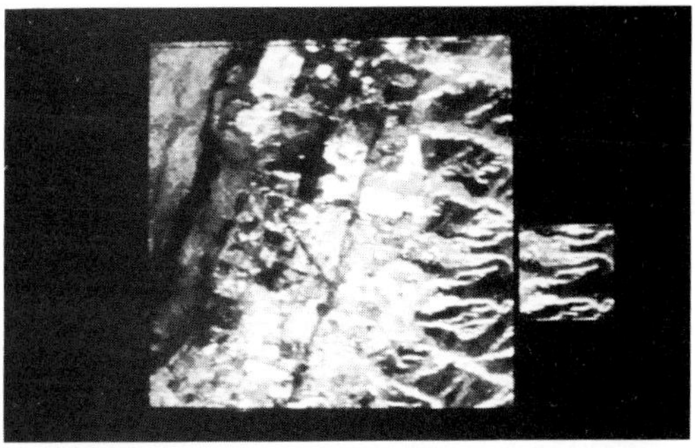

b) Contour map with ΔdB=5.

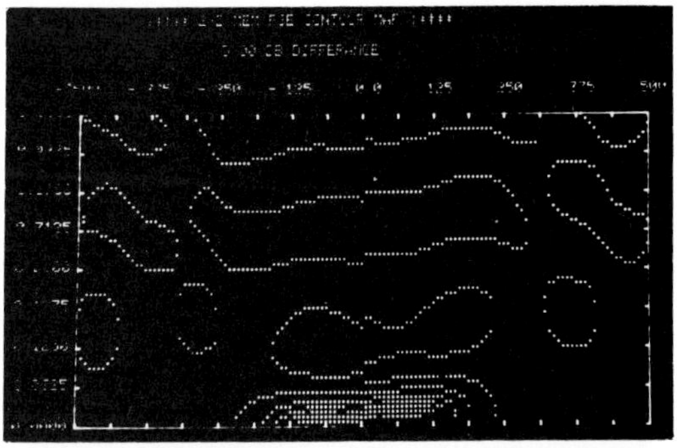

c) 3-D display at (0.0,0.0) point of view

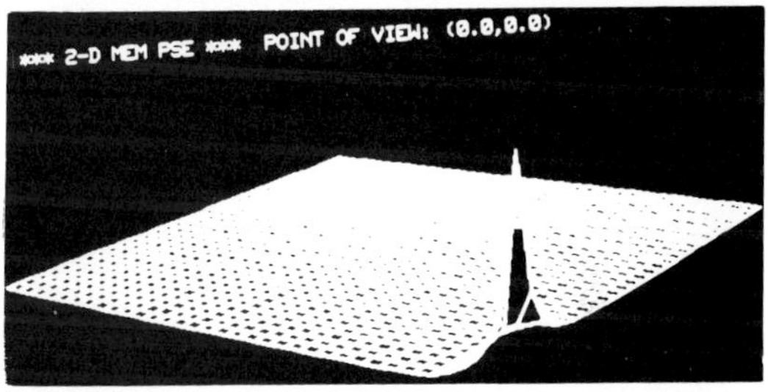

d) 3-D display at (0.5,0.5) point of view

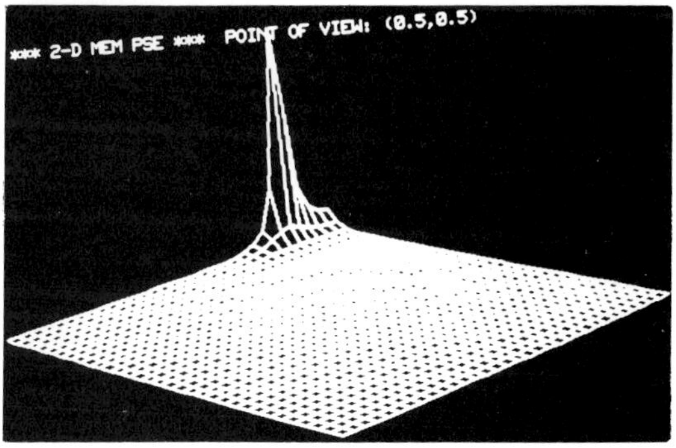

Figure 17

that obtainable by the conventional FFT algorithm. The spectral features suggested for texture discrimination may be the powers associated with the various frequency bands. With its computational efficiency and accuracy, this algorithm is presently the only one available for minicomputer implementation. Further improvement in spectral resolution will make this algorithm even more powerful for the texture-image analysis.

REFERENCES

1. Bajcsy, R., and L. Lieberman, "Texture gradient as a depth cue," Computer Graphics and Image Processing, vol. 5, pp. 52-67, 1976.
2. Weszka, J., D. Dyer and A. Rosenfeld, "A comparative study of texture measures for terrain classification," IEEE Trans. on Systems, Man and Cybernetics, vol. SMC-6, pp. 269-285, 1976.
3. Laws, K., "Texture energy measures," in R. Nevatia and A. Sawchuk, "Semiannual Technical Report," University of Southern California, USCIPI Report 910, pp. 18-31, 1979.
4. Lim, J. S., and N. A. Malik, "A new algorithm for two dimensional maximum entropy power spectral estimation," IEEE Trans. Acoust., Speech, Signal Processing, ASSP-29(3), pp. 401-413, June 1981.
5. Lim, J. S., and N. A. Malik, "Minicomputer implementation of an iterative two-dimensional maximum entropy power spectrum estimation algorithm," Proceedings of the 2nd International Symposium on Computer Aided Seismic Analysis and Discrimination, Aug. 1981.
6. Lim, J. S., and N. A. Malik, "Some properties of two-dimensional maximum entropy power spectrum estimation," First ASSP Workshop on Spectral Estimation, Aug. 1981, vol. 1.
7. Burg, J. P., unpublished notes, 1974.
8. Wornecke, S. J., and L. R. D'Addario, "Maximum entropy image reconstruction," IEEE Trans. Comput., vol. C-26, pp. 351-364, April 1977.
9. Woods, J. W., "Two-dimensional Markov spectral estimation," IEEE Trans. Inform. Theory, vol. IT-22, pp. 552-559, Sept. 1976.
10. Cadzow, J. A., and K. Ogino, "Two dimensional spectral estimation," IEEE Trans. Acoust., Speech, Signal Processing, ASSP-29(3), pp. 396-401, June 1981.

CHAPTER 7

A Study of Texture Classification Using Spectral Features

1. Introduction

To identify objects or region of interest in an image, textural properties are very important because of their useful information about the structural arrangement of surfaces and their relationship to the surrounding environment. Various methods of textural feature selection have been proposed (see e. g. [1][2]). The spectral information has been generally recognized as an effective feature for the texture discrimination and segmentation. A busy picture has relatively high power at high spatial frequencies as compared to a smooth picture. Also, the directional biases in the picture should give rise to the directional biases in the spectrum. And any periodic properties should produce very high values in the power spectrum. The efficient two-dimensional maximum entropy spectral analysis method described in the previous chapter will be used to compute the spectral features for texture classification in this chapter.

2. Feature Selection and Classification Method

We use two features to classify the texture images. It is generally recognized that a coarse texture will have a high value of power spectrum near the origin while in a fine texture, the value will be more spread out. Thus, if one wishes to analyze texture coarseness, a set of features that should be useful are the averages of the power spectrum values taken over a ring-shaped region centered at the origin. Let $\hat{P}_x(\omega_1,\omega_2)$ be the power spectrum estimate of the random process $x(n_1, n_2)$, and $\hat{P}_x(r,\theta)$ be the corresponding spectrum in polar coordinate. Herein we consider only the first quadrant of the power spectrum, i.e.,

$$\phi_r = \int_0^{\frac{\pi}{2}} \hat{P}_x(r, \theta)\, d\theta, \tag{1}$$

for various values of r, the ring radius.

For the discrete case, this can be written as (for rings between radius r_1 and r_2):

$$\phi_{r_1 r_2} = \sum_{r_1^2 \leq x^2+y^2 < r_2^2} \hat{P}_x(x,y) \quad ; \quad 0 \leq x,y \leq n-1 \tag{2}$$

Similarly, it is well known also that the angular distribution of power spectrum is sensitive to the directionality of the texture at frequency $\omega=(\omega_1,\omega_2)$. A texture with many edges or lines in a given direction θ will have high values of power spectrum around the perpendicular of $\theta+\frac{\pi}{2}$; while in a nondirectional texture the spectrum should also be nondirectional. Thus, a good set of features for analyzing the texture directionality should be the averages of the power-spectrum values taken over wedge-shaped regions centered at the origin, i.e.

$$\phi_\theta = \int_0^\infty \hat{P}_x(r,\theta)\, dr \tag{3}$$

for the various values of θ, the wedge slope.

For the discrete case, this is (the wedge between θ_1 and θ_2) given by

$$\phi_{\theta_1\theta_2} = \sum_{\theta_1 \leq \tan^{-1}\frac{y}{x} < \theta_2} \hat{P}_x(x, y) \tag{4}$$

$$o < x, y \leq n-1$$

The features calculated by Eqs. (2) and (4) are sensitive to size and orientation respectively, but not to both. In order to obtain the comparable feature sets, we obtain sets of equalized features by taking the average over the intersection area of rings and wedges. These equalized features are also studied in Section 3 of this chapter.

After the calculation of features, we use the Fisher linear discriminant technique for classification [3]. Let m be the overall mean vector, i.e. the average of all learning samples. For a two-class problem, we decide that a given feature vector x belongs to class 1 (subset H_1), if

$$w^t(x - m) > 0 ; \tag{5}$$

otherwise we decide for class 2 (subset H_2). In Eq. (5), w is defined as

$$w = \alpha n S_w^{-1}(m_1 - m_2) \tag{6}$$

where

α : a scalar

n : number of samples

S_w: the pooled sample scatter matrix = $\sum_{i=1}^{2} \sum_{x \in H_i} (x-m_i)(x-m_i)^t$

m_i: mean = $\frac{1}{n_i} \sum_{x \in H_i} x$, $i=1,2$.

When we have more than two classes, we can use a voting scheme to classify a given vector of measurements, x. For each pair of classes H_i, H_j, we project x on appropriate line and classify it as described above. This gives us k(k-1)/2 different classifications, where k is the number of classes of x. Finally, we assign x to the class that received the majority of votes.

3. Experimental Results

Because of the computational requirements of the method and the limited memory capacity of the PDP 11/45, all test samples are stored in our DEC 20 system and sent through a high speed communication line to the PDP 11/45 for the spectrum computation. The test samples are the texture images taken from the U.S.C. data base. To verify the sensitivity both in coarseness and directionality, we select some textures that contain such information. The test samples come from six classes of texture (each has four samples) and are shown in Fig. 1. Each data is a 32x32 array of gray level 0 - 255. These pictures reappear but are two times larger in Fig. 2(a) - Fig. 7(a). Fig. 2(b) - Fig. 7(b) display the corresponding estimated power-spectrum of the upper left data in each class. We can see that those spectra are different either in radial or angular distribution.

The feature sets we used are:

ring: $\phi_{r_1 r_2}$ for (r_1, r_2) = (1,3), (3,6), (6,12), (12,24), (24,48)

wedge: $\phi_{\theta_1 \theta_2}$ for (θ_1, θ_2) = (0,15), (15,30), (30,45), (45,60), (60,75), (75,90)

The maximum ring radius used is 48 since it already covers almost the whole 64x64 array power spectrum.

A combined feature of ring and wedge has been tested for 30 pairs of feature values. Table I shows a subset of features which did better than 19 out of 24 correct, i.e. more than 80% correct. Table II shows the best performing pairs using the same kind of features (ring and ring, wedge and wedge): there are 6 out of 25 pairs which classified correctly at a rate better than 75%. Other pairs' results are concentrated near 12-17, i.e., more than 50% correct recognition. For the pairs that contain the wedge near the edges, the results are very good since the test samples give some directional information. Also for the rings a little farther from the origin, the results are better since it shows a large difference in the spectrum value there.

Equalized features are also tested: we used five rings intersected with three wedges (ring: (1,3),(3,6),(6,12),(12,24),(24,48) and wedge: (0,30),(30,60),(60,90)). One hundred and five pairs of features have been tested. Table III shows the best performing pairs of which the best score, 23 out of 24, is 95% correct. From the results, we can see that the ring feature (24,48) gives very useful classification information indicating that there exists a large textural variation in that region as the texture coarseness plays an important role in the pair. Fig. 8 is a histogram of scores obtained of all the 105 pairs of test features and indicates a high probability for the overall classification accuracy of the spectral features.

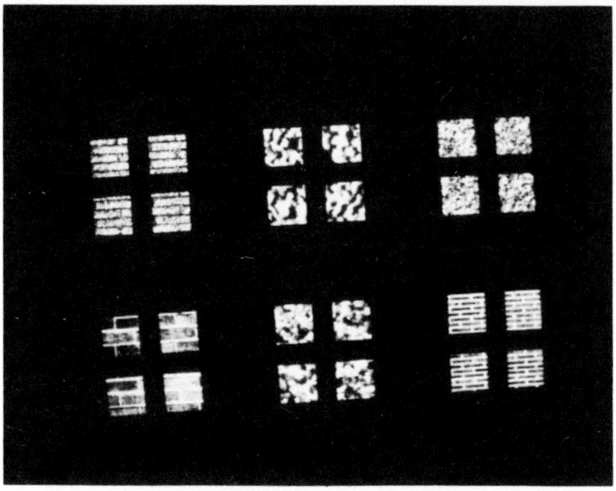

Figure 1 The original test texture samples with 6 classes, 24 samples, and each data format of 32x32.

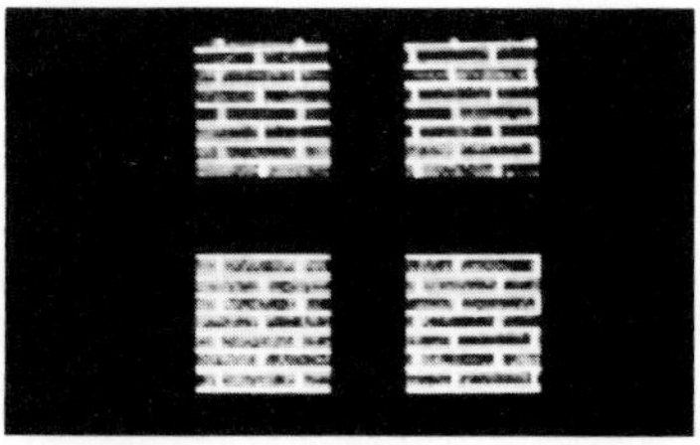

a) Test samples of class 1

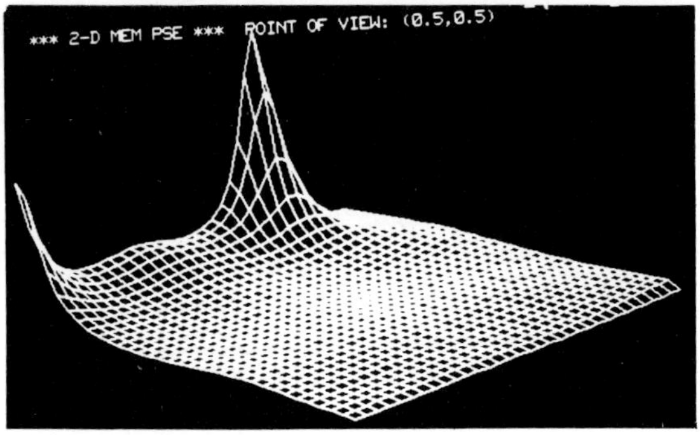

b) The estimated power spectrum for the upper left sample of Fig. 2a

Figure 2

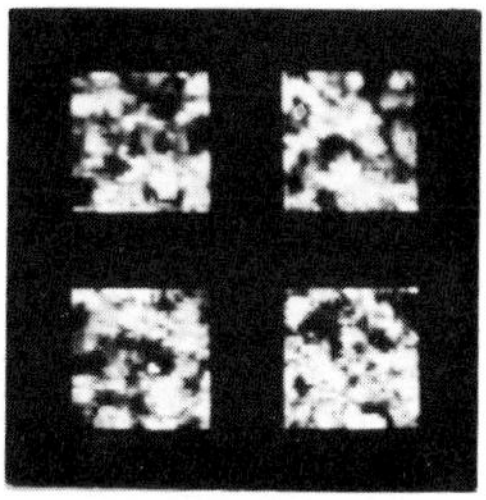

a) Test samples of class 2

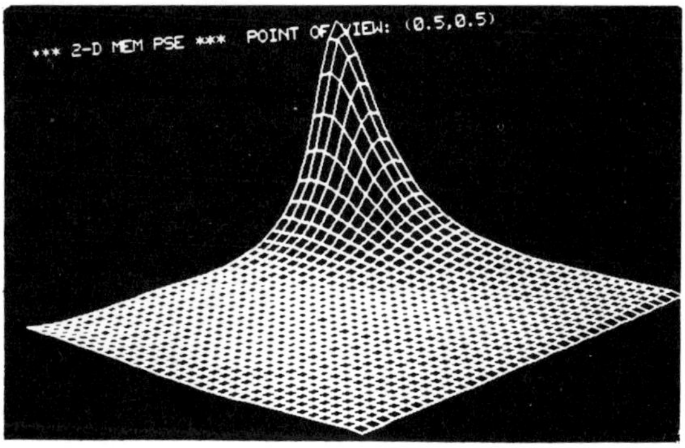

b) The estimated power spectrum for the upper left sample of Fig. 3a

Figure 3

a) Test samples of class 3

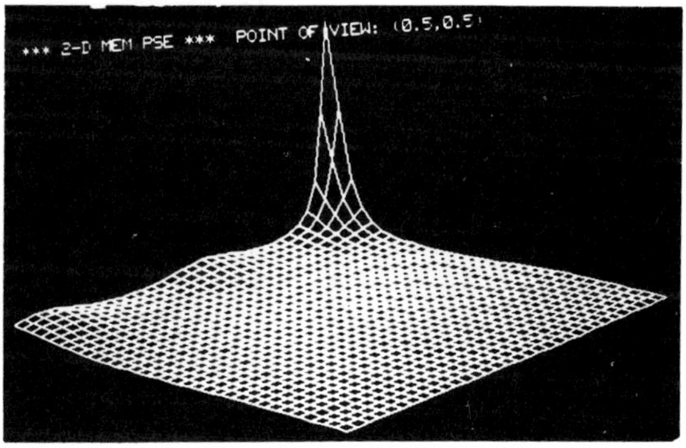

b) The estimated power spectrum for the upper left sample of Fig. 4a

Figure 4

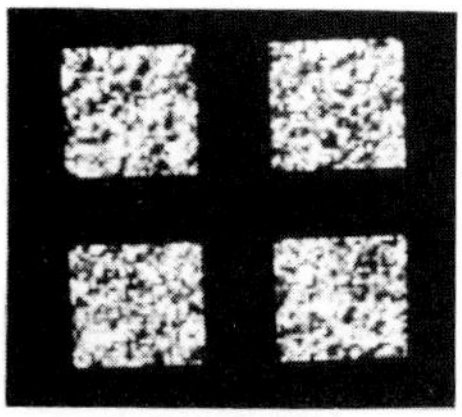

a) Test samples of class 4

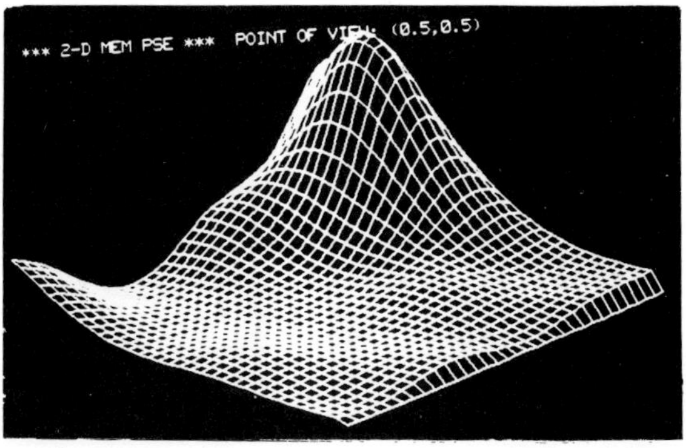

b) The estimated power spectrum for the upper left sample of Fig. 5a

Figure 5

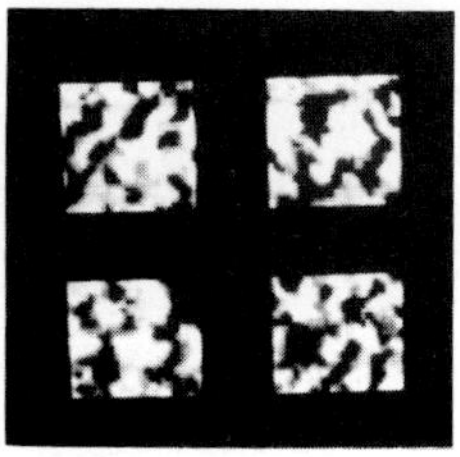

a) Test samples of class 5

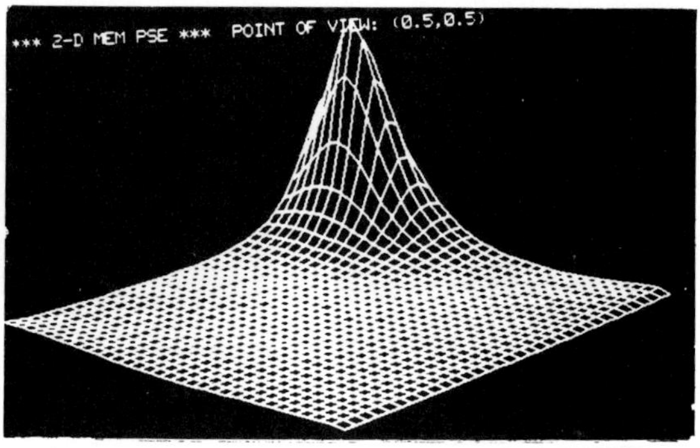

b) The estimated power spectrum for the upper left sample of Fig. 6a

Figure 6

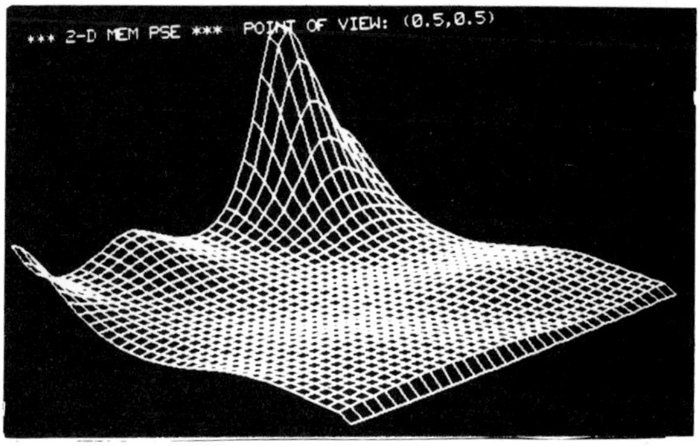

a) Test samples of class 6 b) The estimated power spectrum for the upper left sample of Fig. 7a

Figure 7

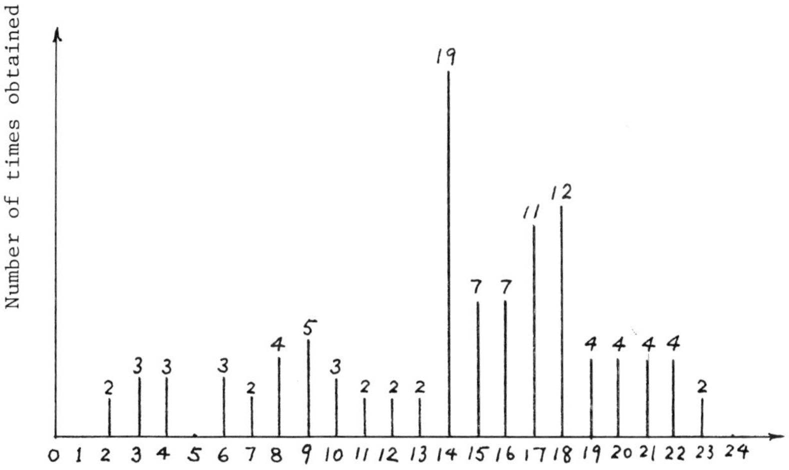

Number of correctly classified samples

Figure 8 Histogram of scores obtained using pairs of equalized features.

Features		Number Correctly Classified
Ring	Wedge	
(24,48)	(0,15)	22
(1,3)	(0,15)	21
(3,6)	(0,15)	20
(1,3)	(45,60)	19
(12,24)	(0,15)	19
(3,6)	(30,45)	19
(3,6)	(75,90)	19

TABLE I Best Performing Pairs Using the Combination Feature of Ring and Wedge for Those with More Than 80% Correct Classification.

Features		Number Correctly Classified
Ring	Ring	
(6,12)	(24,48)	20
(6,12)	(12,24)	19
Wedge	Wedge	
(30,45)	(75,90)	20
(15,30)	(60,75)	18
(30,45)	(60,75)	18
(45,60)	(60,75)	18

TABLE II Best Performing Pairs Using Same Kind of Features, for Those with More Than 75% Correct Classification.

Features				Number Correctly Classified
Ring ∩	Wedge	Ring ∩	Wedge	
(3,6)	(0,30)	(24,48)	(0,30)	23
(12,24)	(60,90)	(24,48)	(0,30)	23
(24,48)	(0,30)	(24,48)	(30,60)	22
(1,3)	(0,30)	(24,48)	(0,30)	22
(1,3)	(30,60)	(24,48)	(0,30)	22
(1,3)	(60,90)	(24,48)	(0,30)	22
(3,6)	(30,60)	(24,48)	(0,30)	21
(3,6)	(60,90)	(24,48)	(0,30)	21
(6,12)	(0,30)	(24,48)	(0,30)	21
(12,24)	(30,60)	(24,48)	(0,30)	21
(12,24)	(0,30)	(12,24)	(60,90)	20
(1,3)	(30,60)	(6,12)	(0,30)	20
(1,3)	(60,90)	(6,12)	(0,30)	20
(12,24)	(0,30)	(24,48)	(0,30)	20
(1,3)	(0,30)	(6,12)	(0,30)	19
(3,6)	(0,30)	(6,12)	(0,30)	19
(3,6)	(30,60)	(6,12)	(0,30)	19
(6,12)	(0,30)	(12,24)	(60,90)	19

TABLE III Best Performing Pairs Using Equalized Features for Those with More Than 80% Correct Classification.

4. Discussion

In this chapter we have observed that equalized features did better than unequalized ones for this set of test samples. It is verified that both the coarseness and the directionality are important factors in texture discrimination. For the consideration of practical use in automatic classification, various kinds of textures must be tested and compared with other methods using the non-spectral features. Another important factor which may influence the results is that if we increase the autocorrelation function and the discrete Fourier transform length while estimating the power spectrum, the accuracy and the resolution will be better. But there is a tradeoff between the accuracy and the computational time. In this chapter these parameters (i.e. autocorrelation function: 7x7, discrete Fourier transform length: 32) are chosen to facilitate real-time processing. Also, the locations of the main and second components of frequencies can serve as other important features because they vary among different textures.

Finally, it should be pointed out that the classification results described in this chapter are quite comparable to those reported by Weszka, et. al. [4] who use co-occurrence matrix for texture feature computation. The spectral features, however, can be computed more efficiently.

REFERENCES

1. Haralick, R. M., K. Shanmugam and I. Dinstein, "Textural features for image classification," IEEE Trans. on Systems, Man and Cybernetics, vol. SMC-3, pp. 610-621, Nov. 1973.
2. Chen, C. H., and P. C. Chen, "A comparative evaluation of texture measures," Technical Note No. EE-75-9, SMU, N. Dartmouth, MA, Sept. 25, 1975.
3. Duda, R. O., and P. E. Hart, "Pattern Classification and Scene Analysis", John Wiley and Sons, New York, 1972.
4. Weszka, J. S., C. Dyer and A. Rosenfeld, "A comparative study of texture measures for terrain classification", IEEE Trans. on Systems, Man and Cybernetics, vol. SMC-6, pp. 269-285, 1976.

CHAPTER 8

Bibliography

I. Introduction

In recent years there has been strong research interest on theory, computational algorithms, and applications of the nonlinear methods of spectral analysis using maximum entropy and related concepts. This area has not reached maturity and many papers and reports will be published in the near future. The activity in signal recognition using the results of the maximum entropy spectral analysis will certainly be increasing also especially in various applications. Thus, it is not possible to provide an exhaustive list of publications in the area. This Bibliography, however, will serve as a useful reference for anyone working in this area.

In addition to regular publications, such as IEEE ASSP Transactions and annual Conference Proceedings, Signal Processing, Geophysics, etc., the Proceedings of the 1978 and 1979 RADC Spectrum Estimation Workshop, and the Proceedings of the First ASSP Workshop on Spectral Estimation, held at McMaster University, Hamilton, Ontario, Canada, Aug. 17-18, 1981, contain a number of papers in this area. In this chapter we present not only the bibliography on the maximum entropy methods in one and two spatial dimensions, but also a number of related methods of high resolution spectral analysis. One common assumption with all these methods is that the data record is short and thus the conventional fast Fourier transform method of spectral analysis is not suitable. In the following sections, references are arranged in the first author's alphabetical order. Each reference is listed only once in the chapter.

II. Bibliography on Univariate Maximum Entropy Spectral Analysis

1. Ables, J. G., "Maximum entropy spectral analysis", Astron. Astrophysics Supplement, vol. 15, pp. 383-393, 1974.
2. Akaike, H., "Power spectrum estimation through autoregressive model filtering", Annals of Inst. Stat. Math., vol. 21, pp. 407-419, 1969.
3. Akaike, H., "A new look at the statistical model identification", IEEE Trans. on Automatic Control, vol. AC-19, pp. 716-723, 1974.

4. Andersen, N., "On the calculation of filter coefficients for maximum entropy spectral analysis", Geophysics, vol. 39, no. 1, pp. 69-72, 1974.
5. Baggeroer, A. B., "Confidence intervals for regression (MEM) spectral estimates", IEEE Trans. on Information Theory, vol. IT-22, no. 5, pp. 534-545, 1976.
6. Barnard, T. and J. P. Burg, "Analytical studies of techniques for the computation of high-resolution wavenumber spectra," Texas Instruments, Inc., Adv. Array Res. Rep. #9 (AD 855345) May 1969.
7. Bowling, S. B., "Linear prediction and maximum entropy spectral analysis for radar applications", MIT Lincoln Lab. Project Report RMP 122 (ESD-TR-113), 1977.
8. Burg, J. P., "Maximum entropy spectral analysis", Proc. of the 37th Meeting of the Society of Exploration Geophysicists, 1967.
9. Burg, J. P., "A new analysis technique for time series data", presented at the NATO Advanced Study Institute on Signal Processing with Emphasis on Underwater Acoustics, Enschede, the Netherlands, Aug. 1968.
10. Burg, J. P., "The relationship between maximum entropy spectra and maximum likelihood spectra", Geophysics, vol. 37, pp. 375-376, April 1972.
11. Burg, J. P. "Maximum entropy spectral analysis", Ph.D. Thesis, Dept. of Geophysics, Stanford University, Stanford, CA, 1975.
12. Chan, H. C. and S. Haykin, "Applications of the maximum entropy method in radar signal processing", Report CRL-62, Communications Research Lab., Faculty of Engineering, McMaster University, Hamilton, Canada, March 1979.
13. Chen, C. H., "A non-linear maximum entropy method for spectral estimation", Final Report prepared for SCEEE, August 1979.
14. Chen, C. H., et. al., "A minicomputer implementation of Fougere's maximum entropy spectrum analysis method", Report for AFOSR Project no. NP-80-018, SMU-EE-TR-80-7, August 1980.
15. Chen, W. Y. and G. R. Stegen, "Experiments with maximum entropy power spectra of sinusoids", Journal of Geophysics Research, vol. 79, no. 20, PP. 3019-3022, July 1974.
16. Childers, D. G. (Editor), "Modern Spectrum Analysis", IEEE Press, New York 1978.
17. Dyson, T. and S. S. Rao, "Equal observation interval comparison of maximum entropy and weighted overlapped segment averaging spectrum estimation techniques", IEEE ASSP Transactions, vol. 29, no. 4, pp. 919-923, August 1981.
18. Dowla, U. and J. H. McClellan, "MEM spectral analysis for non-uniformly sampled signals," Proc. of the Second International Symposium on Computer-aided Seismic and Discrimination, August, 1981.
19. Fougere, P. F., "On nonlinear maximum entropy power spectrum analysis", Proc. of the Second International Symposium on Computer-aided Seismic Analysis and Discrimination, August 1981.
20. Fougere, P. F., "A solution to the problem of spontaneous line splitting in maximum entropy power spectrum analysis", Journal of Geophysical Research, vol. 82, p. 1051, 1977.
21. Fougere, P. F., E. J. Zawalick and H. R. Radoski, "Spontaneous line splitting in maximum entropy power spectrum analysis", Physics of Earth and Planetary Interiors, vol. 12, p. 201, 1976.

22. Fougere, P. F., "Sunspots: power spectra and a forecast", International Solar-Terrestrial Predictions Proceedings and Workshop Program, Preprint no. 103, December 1978.
23. Fougere, P. F., "A solution to the problem of spontaneous line splitting in maximum entropy power spectrum analysis of complex signals", Proc. of 1978 RADC Spectrum Estimation Workshop, pp. 77-84.
24. Gabriel, W. F., "Antenna spatial pattern viewpoint of MEM, MLM and adaptive array resolution", Proc. of 1979 RADC Spectrum Estimation Workshop.
25. Gersch, W. and D. R. Sharpe, "Estimation of power spectra with finite order autoregressive models", IEEE Trans. on Automatic Control, vol. AC-18, pp. 367-369, 1973.
26. Haykin, S., "Spectral classification of radar clutter using the maximum entropy method", in "Pattern Recognition and Signal Processing", edited by C. H. Chen, Sijthoff and Noordhoff, 1978.
27. Haykin S. (editor) "Nonlinear Methods of Spectral Analysis", Springer-Verlag, 1979.
28. Haykin, S., S. Kesler and B. Currie, "An experimental classification of radar clutter", Proc. of IEEE, vol. 67, pp. 332-333, February 1979.
29. Haykin, S. and S. Kesler, "The complex form of the maximum entropy method for spectral estimation", Proc. of IEEE, vol. 64, no. 5, pp. 822-823, May 1976.
30. Herring, R. W., "A comparison of the Burg and the known-autocorrelation autoregressive spectral analysis of complex sinusoidal signals in additive white noise", Proc. of the 1979 RADC Spectrum Estimation Workshop. Also as Report AD A080020.
31. Herring, R. W., "Maximum entropy spectral analysis and radar signal processing", CRC Report no. 1330, Communications Research Centre, Dept. of Communications, Ottawa, Canada, January 1980. Also as Report AD A01343.
32. Hsu, M., "Maximum entropy principle and its application to spectral analysis and image reconstruction", Ph.D. thesis, Ohio State University, 1975.
33. Hung, E. K. L. and R. W. Herring, "Simulation experiments to compare the signal detection properties of DFT and MEM spectra,"IEEE ASSP Transactions, vol. 29, no. 5, pp. 1084-1089, October 1981.
34. Jones, R. H., "Identification and autoregressive spectrum estimation", IEEE Trans. on Automatic Control, vol. AC-19, pp. 894-897, December 1974.
35. Jones, R. H., "Autoregressive order selection", Geophysics, vol. 41, pp. 771-773, 1976.
36. Kanasewich, E. R., "Time Series Analysis in Geophysics", University of Alberta Press, Edmonton, Canada, 1973.
37. Kaveh, M. and G. R. Cooper, "An empirical investigation of the properties of the autoregressive spectral estimator", IEEE Trans. on Information Theory, vol. IT-22, pp. 313-323, May 1976.
38. Kay, S., "Detection of a sinusoid in white noise by autoregressive spectrum analysis", Proc. of 1980 IEEE ASSP Conference, pp. 658-661.
39. Kesler, S. B. and S. Haykin, "The maximum entropy method applied to the spectral analysis of radar clutter", IEEE Trans. on Information Theory, vol. IT-24, no. 2, pp. 269-272, March 1978.

40. Keeler, R. J., "Uncertainties in adaptive maximum entropy frequency estimators", IEEE Trans. on ASSP, vol. 26, pp. 469-471, Oct. 1978.
41. King, W. R., "Applications for MESA and the prediction error filter", Proc. of 1979 RADC Spectrum Estimation Workshop.
42. Lacoss, R. T., "Data adaptive spectral analysis methods", Geophysics, vol. 36, pp. 661-675, 1971.
43. Landers, T. E. and R. T. Lacoss, "Some geophysical applications of autoregressive spectral estimates", IEEE Trans. on Geoscience Electronics, vol. GE-15, pp. 26-32, January 1977.
44. Lang, S. W., "Near optimal frequency/angle of arrival estimates based on maximum entropy spectral techniques", Proc. of 1980 IEEE ICASSP, pp. 829-832.
45. Marple, L., "A new autoregressive spectrum analysis algorithm", Proc. of the 1979 RADC Spectrum Estimation Workshop, pp. 51-58.
46. Marple, L., "Frequency resolution of high-resolution spectrum analysis techniques", Proc. of the 1978 RADC Spectrum Estimation Workshop, pp. 19-35.
47. Moorcroft, D. R., "Maximum entropy spectral analysis of radio-auroral signals", Radio Science, vol. 13, pp. 649-660, July-August 1978.
48. Johnsen, S. J. and N. Andersen, "On power estimation in maximum entropy spectral analysis", Geophysics, vol. 43, pp. 681-690, June 1978.
49. Newman, W. I., "Extension to the maximum entropy method", IEEE Trans. on Information Theory, Part I in vol. IT-23, pp. 89-93, 1977; Part II in vol. IT-25, no. 6, pp. 705-708, November 1979.
50. Nuttall, A. H., "Spectral analysis of univariate process with bad data points, via maximum entropy, and linear predictive techniques", Naval Underwater Systems Center, TR no. 5303, 1976.
51. Papoulis, A., "Signal Analysis", McGraw-Hill Company, pp. 248-251, 1977.
52. Papoulis, A., "Maximum entropy and spectral estimation: a review," IEEE ASSP Transactions, vol. 29, no. 6, pp. 1176-1186, December 1981.
53. Radoski, H. R., P. F. Fougere, and E. J. Zawalick, "A comparison of power spectral estimates and applications of the maximum entropy method", Journal of Geophysical Research, vol. 80, p. 619, 1975.
54. Radoski, H. R., E. J. Zawalick and P. F. Fougere, "The superiority of maximum entropy power spectrum techniques applied to geomagnetic micropulsations", Physics of Earth and Planetary Interiors, vol. 12, p. 208, 1976.
55. Reilly, J. and S. Haykin, "An experimental study of the MEM applied to array antennas in the presence of multipath", Proc. of 1980 IEEE ICASSP, pp. 120-123.
56. Robinson, E. A. and S. Treitel, "Maximum entropy and the relationships of the partial autocorrelation to the reflection coefficients of a layered system", IEEE Trans. on Acoustic, Speech and Signal Processing, vol. 28, no. 2, pp. 224-235, April 1980.
57. Satorius, E. H. and J. R. Zeidler, "Maximum entropy spectral analysis of multiple sinusoids in noise", Technical Report 201, Naval Ocean Systems Center, San Diego, CA, March 1977. Also as report AD A053257.

58. Sawyers, J. H., "The maximum entropy method applied to radar adaptive doppler filtering", Proc. of 1979 RADC Spectrum Estimation Workshop, pp. 289-301.
59. Smylie, D. E., G. K. C. Clarke and T. J. Ulrych, "Analysis of irregularities in the Earth's rotation", in "Methods in Computational Physics," vol. 13, Academic Press, pp. 391-430, 1973.
60. Steen, A., C. H. Chen, and G. Wannberg, "A nonlinear maximum entropy method for spectral estimation applied to incoherent scatter radar measurements", submitted to publication, 1982.
61. Swingler, D. N., "A comparison between Burg's maximum entropy method and a nonrecursive technique for the spectral analysis of deterministic signals", Journal of Geophysical Research, vol. 84, pp. 679-685, Feb. 1979.
62. Theodorides, S. and D. C. Cooper, "Application of the maximum entropy spectrum analysis technique to signals with spectral peaks of finite width," Signal Processing Journal, vol. 3, pp. 109-122, 1981.
63. Toomey, J. P., "High resolution frequency measurement by linear prediction", vol. AES-16, no. 4, pp. 517-525, July 1980.
64. Ulrych, T. J. and T. N. Bishop, "Maximum entropy spectral analysis and autoregressive decomposition", Review of Geophysics and Space Physics, vol. 33, pp. 183-200, 1975.
65. Ulrych, T. J. and R. W. Clayton, "Time series modelling and maximum entropy", Physics of the Earth and Planetary Interior, vol. 12, pp. 188-200, 1976.
66. Ulrych, T. J., "Maximum entropy power spectrum of long period geomagnetic reversals", Nature, vol. 235, pp. 218, 1972.
67. van den Bos, A., "Alternative interpretation of maximum entropy spectral analysis", IEEE Trans. on Information Theory, vol. IT-17, pp. 493-494, 1971.
68. Zeidler, J. R., et. al., "Adaptive enhancement of multiple sinusoids in uncorrelated noise", IEEE Trans. on ASSP, vol. 26, pp. 240-254, June 1978.

III. Bibliography on Multichannel (Multivariate) Maximum Entropy Spectral Analysis and Array Processing

1. Chen, C. H. and C. Yen, "On multichannel (multivariate) maximum entropy spectral analysis", Technical Report, SMU-EE-TR-81-2, January 1981.
2. Fougere, P. F., "Spontaneous line splitting in multichannel maximum entropy power spectra", Proc. of the First ASSP Workshop on Spectral Estimation, August 1981.
3. Ioannidis, G. A., "Application of multivariate autoregressive spectrum estimation to ULF waves", Radio Science, vol. 10, no. 12, pp. 1043-1054, 1975.
4. Jones, R. H., "Multivariate autoregression estimation using residuals", in "Applied Time Series Analysis," D. F. Findley, ed., Academic Press, 1978.
5. Lee, T. S., "Identification and spectral estimation of noisy multivariate autoregressive processes", Proc. of ICASSP, 1981.
6. McDonough, R. N., "Maximum entropy spatial processing of array data", Geophysics, vol. 39, no. 6, pp. 843-851, December 1974.

7. Morf, M., A. Vieira, D. T. L. Lee and T. Kailath, "Recursive multichannel maximum entropy spectral estimation", IEEE Trans. on Geoscience Electronics, vol. GE-16, no. 2, pp. 85-94, April 1978.
8. Nuttall, A. H., "Multivariate linear predictive spectral analysis employing weighted forward and backward averaging: a generalization of Burg's algorithm", NUSC-TR-5501, 1976.
9. Nuttall, A. H., "Fortran program for multivariate linear predictive spectral analysis employing forward and backward averaging", NUSC-TR-5419, 1976.
10. Nuttall, A. H., "Positive definite spectral estimate and stable correlation recursion for multivariate linear predictive spectral analysis", NUSC-TR-5729, 1977.
11. Parzen, E., "Multiple time series modelling", in "Multivariate Analysis," vol. 2, P. R. Krishnaian, Ed., Academic Press, New York, pp. 389-409, 1969.
12. Robinson, E. A., "Multichannel Time Series Analysis with Digital Computer Programs", Holden-Day, San Francisco, 1967.
13. Skinner, D. P., S. M. Hedlicka and A. D. Matthews, "Maximum entropy array processing", Journal of Acoustical Society of America, vol. 66, no. 2, August 1979.
14. Strand, O. N., "Multichannel complex maximum entropy (autoregressive) spectral analysis", IEEE Trans. on Automatic Control, vol. AC-22, no. 4, pp. 634-640, August 1977.
15. Therrien, C. W., "Relations between 2-D and multichannel linear prediction", IEEE ASSP Transactions, vol. 29, no. 3, pp. 454-456, June 1981.
16. Tyraskis, P. A. and O. G. Jensen, "Multichannel autoregressive data models," presented at the 42nd EAEG Meeting in Taksim-Istanbul, 1980.
17. Wiggins, R. A. and E. A. Robinson, "Recursive solution to the multichannel filtering problem", Journal of Geophysical Research, vol. 70, no. 8, pp. 1885-1891, 1965.

IV. Bibliography on Related High Resolution Spectral Analysis

1. Cadzow, J. A., "High performance spectral estimation - a new ARMA method", IEEE Trans. on ASSP, vol. 28, no. 5, pp. 524-529, October 1980.
2. Cadzow, J. A., "Improved spectral estimation from incomplete sampled-data observations", Proc. of 1978 RADC Spectrum Estimation Workshop.
3. Cadzow, J. A. and R. Moses, "A superresolution method of ARMA spectral estimation", Proc. of ICASSP, 1981.
4. Cadzow, J. A., "ARMA modeling of time series," IEEE Trans. on Pattern Analysis and Machine Intelligence, vol. PAMI-4, no. 2, pp. 124-128, March 1982.
5. Capon, J., "High resolution frequency-wave number spectrum analysis", Proc. of IEEE, vol. 57, pp. 1408-1419, August 1969.
6. Carter, G. C. and A. H. Nuttall, "On the weighted overlapped segment averaging method for power spectral estimation", Proc. of IEEE, October 1980.
7. Clarke, R. H. and D. V. Tibble, "Measurement of the elevation angles of arrivals of multicomponent H. F. skywaves", Proc. of IEEE, vol. 125, pp. 17-24, January 1978.

8. deFigueiredo, R. J. P., "An optimum filter for spectral estimation based on a penality function approach", Proc. of 1978 RADC Spectrum Estimation Workshop.
9. deFigueiredo, R. J. P., "Application of a frequency domain Prony method to wide bandwidth radar signature classification", presented at the 1980 Workshop on Digital Signal and Waveform Analysis, Miami Beach, Florida, December 1980.
10. Evans, J. E. and D. F. Sun, "Aperture sampling processing for ground reflection elevation multipath characterization", Proc. of the 1979 RADC Spectrum Estimation Workshop.
11. Fahlman, G. G. and T. J. Ulrych, "On the light curve of 3C273", Astrophysical Journal, vol. 201, pp. 277-286, October 1975.
12. Frost, O. L., "Power spectrum estimation", in "Aspects of Signal Processing," NATO ASI Proceedings, D. Reidel Publishing Company, Dordrecht-Holland, 1977.
13. Gray, H. L., A. G. Houston and F. W. Morgan, "On G-spectral estimation", in "Applied Time Series Analysis", D. F. Findley, Ed., Academic Press, 1978.
14. Griffiths, L. J. and R. Prieto-Diaz, "Spectral analysis of natural seismic events using autoregressive techniques", IEEE Trans. on Geoscience Electronics, vol. GE-15, no. 1, pp. 13-25, January 1977.
15. Griffiths, L. J., "Rapid measurement of digital instantaneous frequency", IEEE Trans. on ASSP, vol. 23, pp. 207-222, April 1975.
16. Guarino, C. R., "Tone estimation using an autoregressive-moving average model", presented at the 1980 Workshop on Digital Signal and Waveform Analysis, Miami Beach Florida, December 1980.
17. Gutowski, P. R., E. A. Robinson and S. Treitel, "Spectral estimation fact or fiction", IEEE Trans. on Geoscience Electronics, vol. GE-16, no. 22, pp. 8084, April 1978.
18. Jackson, P. L., L. S. Joyce and G. B. Feldkamp, "Application of maximum entropy frequency analysis to synthetic aperture radar", Proc. of the 1978 RADC Spectrum Estimation Workshop.
19. Kay, S. M., "A new ARMA spectral estimator", IEEE Trans. on ASSP, vol. 28, no. 5, pp. 585-588, October 1980.
20. Kaveh, M., "High resolution spectral estimation for noisy signals", IEEE Trans. on ASSP, vol. 27, no. 3, pp. 286-287, June 1979.
21. Makhoul, J., "Lattice methods in spectral estimation", Proc. of 1978 RADC Spectrum Estimation Workshop.
22. Makhoul, J., "Linear prediction: a tutorial review", Proc. of IEEE, vol. 63, pp. 561-580, April 1975.
23. Morf, M. and D. T. Lee, "Recursive spectral estimation of α-stationary processes", Proc. of 1978 RADC Spectrum Estimation Workshop.
24. Nuttall, A. H. and G. C. Carter, "A generalized framework for power spectral estimation", IEEE Trans. on ASSP, vol. 28, no. 3, June 1980. See also the summary that appears in Signal Processing Journal, vol. 2, 1980.
25. Owsley, N. L., "Adaptive data orthogonalization", Proc. of 1978 IEEE ICASSP.

26. Papoulis, A., "A new algorithm in spectral analysis and band-limited extrapolation", IEEE Trans. on Circuits and Systems, vol. CAS-22, no. 9, pp. 735-742, September 1975.
27. Pisarenko, V. F., "The retrieval of harmonics from a covariance function", Geophysics Journal of Royal Astronomical Society, vol. 33, pp. 247-266.
28. Rife, D. C. and R. R. Boorstyn, "Single tone parameter estimation from discrete time observations", IEEE Trans. on Information Theory, vol. IT-20, pp. 591-598, September 1974.
29. Rife, D. C. and R. R. Boorstyn, "Multiple tone parameter estimation from discrete time observations", BSTJ, pp. 1389-1410, November 1976.
30. Stammler, W., "Application of prediction error filters for the detection of weak teleseismic events," IEEE Trans. on Geoscience and Remote Sensing, vol. GE-19, no. 4, October 1981.
31. Tufts, D. W. and R. Kumaresan, "Improved spectral resolution", Part I in Proc. of the IEEE, vol. 68, March 1980; Part II in Proc. of IEEE ICASSP, 1980, pp. 592-597; Part III in Proc. of the IEEE, vol. 68, October 1980.
32. Tufts, D. W., "Adaptive line enhancement and spectrum analysis", Proc. of the IEEE, vol. 65, pp. 169-170, January 1977.
33. van Blaricum, M. L., "A review of Prony's method techniques for parameter estimation", Proc. of the 1978 RADC Spectrum Estimation Workshop.

V. Two-Dimensional Maximum Entropy Spectral Analysis

1. Cadzow, J. A., and K. Ogino, "Two-dimensional spectral estimation", IEEE ASSP Transactions, vol. 29, no. 3, pp. 396-401, June 1981.
2. Dickinson, B. W., "Two-dimensional Markov spectrum estimates need not exist", IEEE Trans. on Information Theory, vol. IT-26, no. 1, pp. 120-121, Jan. 1980.
3. Frieden, B. R., "Image enhancement and restoration", in "Picture Processing and Digital Filtering" ed. by T. S. Huang, Springer-Verlag, 1975.
4. Frieden, B. R., "Problems associated with the maximum entropy image restoration technique", in "Image Science Mathematics", edited by C. O. Wilde and E. Barrett, Western Periodicals Co., 1977.
5. Jain, A. K. and S. Ranganath, "Two dimensional spectral estimation", Proc. of 1978 RADC Spectrum Estimation Workshop.
6. Kikuchi, R. and B. H. Soffer, "Maximum entropy image restoration, Part I: The entropy expression", Journal of Optical Society of America, vol. 67, pp. 1656-1665, December 1977.
7. Lim, J. S. and N. A. Malik, "A new algorithm for two-dimensional maximum entropy power spectrum estimation", vol. 29, no. 3, pp. 401-413, June 1981.
8. Lim, J. S. and N. A. Malik, "Maximum entropy spectrum estimation of signals with missing correlation points", IEEE ASSP Transactions, vol. 29, no. 6, pp. 1215-1217, December 1981.
9. Ong, C., "An investigation of two new high-resolution two dimensional spectral estimate techniques", Texas Instruments, Inc., Long Period Array Processing Report #1, April 1971.

10. Roucos, S. E. and D. G. Childers, "A two-dimensional maximum entropy spectral estimator", IEEE Trans. on Information Theory, vol. IT-26, no. 5, pp. 554-560, September 1980.
11. Wernecke, S. J., "Two dimensional maximum entropy reconstruction of radio brightness", Radio Science, vol. 12, no. 5, pp. 831-844, September/October 1977.
12. Wernecke, S. J. and L. R. D'Addario, "Maximum entropy image reconstruction", IEEE Trans. on Computers, vol. C-26, pp. 351-364, April 1977.
13. Woods, J. W., "Two-dimensional Markov spectral estimation", IEEE Trans. on Information Theory, vol. IT-22, pp. 552-559, September 1976.

APPENDIX A

Flow Diagram of the Nonlinear Maximum Entropy Spectral Analysis

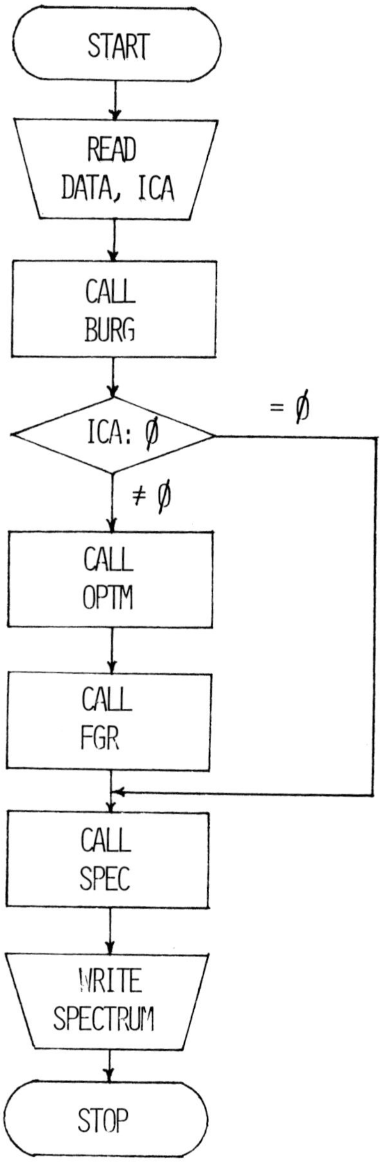

Functional Description

BURG: compute the prediction error filter coefficients based on the Burg's maximum entropy spectral analysis.

OPTM: to find a local minimum of a function of several variables by the method of Fletcher and Powell (a rapid descent method for minimization).

FGR: to find the diagonal elements of the prediction error coefficient matrix, mean square prediction error, and the gradient vector of mean square prediction error with respect to the angles.

SPEC: to find the power spectrum from the prediction error coefficients.

APPENDIX B

Computer Program for Nonlinear Complex Signal Maximum Entropy Spectral Analysis

```
C
C     NAME: OPTM.FOR
C     DATE: AUG. 12, 1981
C
      DOUBLE PRECISION XR(25),XI(25),U,PM,POWER,EXTA(820)
      REAL XB(251),CAL(251)
      COMMON /BLKO/LX,LA,U,PM,POWER/BLK1/XR,XI/BLK2/EXTA
      DATA NF/251/
      DEFINE FILE 1(2,502,U,LINE)
      DO 9000 IIII=1,2
      ACCEPT 10,LX,LA,ICA,LIMIT,LG,DT,PH1,PH2,F1,F2,A1,A2,SNR,U
10    FORMAT(5I10,8F12.5,D15.8)
      RNM=GC(SNR)
      DO 20 I=1,NF
20    XB(I)=FLOAT(I)
      CALL INPUT(F1,F2,PH1,PH2,A1,A2,DT,RNM)
      DO 30 I=1,LX
30    CAL(I)=XR(I)
      CALL KBPLOT(XB,CAL,LX,10,1000,400,700)
      DO 40 I=1,LX
40    CAL(I)=XI(I)
      CALL KBPLOT(XB,CAL,LX,10,1000,50,350)
      CALL BELL
      ACCEPT 50,NX
50    FORMAT(10I10)
      CALL BURG(ICA)
      IF(LG.GE.0)CALL MESG
      IF(ICA.EQ.0)GO TO 60
      CALL OPTM(LIMIT,LG)
      CALL FGR(1)
      IF(LG.GE.0)CALL MESG
60    CALL SPEC(DT,NF)
      DO 70 I=1,NF
70    CAL(I)=EXTA(500+I)
      CALL KBPLOT(XB,CAL,NF,10,1000,100,700)
      IF(ICA.EQ.0)LINE=1
      IF(ICA.NE.0)LINE=2
      WRITE(1'LINE)(CAL(KKK),KKK=1,NF)
      DO 80 I=1,NF
      CAL(I)=0.
80    IF(MOD(I,25).EQ.1)CAL(I)=1.
      CAL(1)=40.
      CALL KBPLOT(XB,CAL,NF,10,1000,100,700)
      CALL BELL
9000  CONTINUE
      CALL EXIT
      END
```

```
      SUBROUTINE INPUT(F1,F2,PH1,PH2,A1,A2,DT,RNM)
      DOUBLE PRECISION XR(25),XI(25),U,PM,POWER
      COMMON /BLKO/LX,LA,U,PM,POWER/BLK1/XR,XI
      PI=ATAN(1.)*4.
      W1=2.*PI*F1
      W2=2.*PI*F2
      PH1=PH1/180.*PI
      PH2=PH2/180.*PI
      POWER=0.DO
      T=0.
      DO 10 I=1,LX
      XR(I)=A1*COS(W1*T+PH1)+A2*COS(W2*T+PH2)+RNM*GAUSS(0.,1.)
      XI(I)=A1*SIN(W1*T+PH1)+A2*SIN(W2*T+PH2)+RNM*GAUSS(0.,1.)
      POWER=POWER+(XR(I)*XR(I)+XI(I)*XI(I))
   10 T=T+DT
      POWER=POWER/DFLOAT(LX)
      RETURN
      END
      FUNCTION GC(SNR)
      VA=SNR/10.
      VB=1./(2.*(10.**VA))
      GC=SQRT(VB)
      RETURN
      END
      FUNCTION GAUSS(XBAR,SIGMA)
      ROOT=1./SQRT(8.*ATAN(1.))
   10 CONTINUE
      X=10.*(RAN(0,1)-.5)
      Y=ROOT*EXP(-X*X/2.)
      YTRY=ROOT*RAN(0,1)
      IF(YTRY.GT.Y)GO TO 10
      GAUSS=X*SIGMA+XBAR
      RETURN
      END
      SUBROUTINE BURG(ICA)
      COMPLEX G(10),PEF(25),PER(25),H(10),F(25),GGG(10),P,Q,SN,SD
      DOUBLE PRECISION XR(25),XI(25),AR(10),AI(10),GR(10,10)
     1,GI(10,10),U,PM,POWER,THETA(20),TMPR,TMP,EXTA(760)
      COMMON /BLKO/NUM,LA,U,PM,POWER/BLK1/XR,XI/BLK2/EXTA
      EQUIVALENCE (EXTA(1),GR(1,1)),(EXTA(101),GI(1,1)),
     1(EXTA(201),AR(1)),(EXTA(211),AI(1)),(EXTA(221),THETA(1))
     2,(EXTA(261),PEF(1)),(EXTA(301),PER(1)),(EXTA(341),F(1)),
     3(EXTA(381),H(1)),(EXTA(391),G(1)),(EXTA(401),GGG(1))
      PM=POWER
      DO 10 I=1,NUM
   10 F(I)=CMPLX(SNGL(XR(I)),SNGL(XI(I)))
      DO 110 NN=1,LA
      XERO=(0.,0.)
      N=NN-1
      IF(N.NE.0)GO TO 30
      DO 20 J=1,NUM
      PEF(J)=XERO
   20 PER(J)=XERO
   30 SN=XERO
```

```
      SD=XERO
      JJ=NUM-N-1
      DO 40 J=1,JJ
      Q=F(J+N+1)+PEF(J)
      P=F(J)+PER(J)
      SN=SN+CONJG(P)*Q
      SD=SD+P*CONJG(P)+Q*CONJG(Q)
40    CONTINUE
      GGG(NN)=-2.*SN/SD
      GR(NN,NN)=REAL(GGG(NN))
      GI(NN,NN)=AIMAG(GGG(NN))
      TMPR=CABS(GGG(NN))
      IF(TMPR.GE.1.)TMPR=.99999
      IF(ICA.EQ.0)GO TO 50
      TMP=TMPR/U
      THETA(NN)=DATAN(TMP/DSQRT(1.DO-TMP*TMP))
      THETA(NN+LA)=DATAN2(GI(NN,NN),GR(NN,NN))
50    PM=PM*(1.-TMPR*TMPR)
      IF(N.EQ.0)GO TO 80
      DO 60 J=1,N
      K=N-J+1
      H(J)=GGG(J)+GGG(NN)*CONJG(GGG(K))
60    CONTINUE
      DO 70 J=1,N
70    GGG(J)=H(J)
      JJ=JJ-1
80    DO 90 J=1,JJ
      PER(J)=PER(J)+CONJG(GGG(NN))*(PEF(J)+F(J+NN))
      PEF(J)=PEF(J+1)+GGG(NN)*(PER(J+1)+F(J+1))
90    CONTINUE
      DO 100 I=1,NN
100   G(I)=GGG(I)
110   CONTINUE
      DO 120 I=1,LA
      AR(I)=REAL(G(I))
120   AI(I)=AIMAG(G(I))
      RETURN
      END
      SUBROUTINE SPEC(DT,NF)
      DOUBLE PRECISION AR(10),AI(10),PM,DT,ANG,U,EXTS,TPI,EXTA(760)
     1,F,FINC,SCSR,SCSI
      COMMON /BLKO/LX,M,U,PM,EXTS/BLK2/EXTA
      EQUIVALENCE (EXTA(201),AR(1)),(EXTA(211),AI(1))
      TPI=8.DO*DATAN(1.DO)
      F=-1./4./DT
      FINC=1./DT/((NF-1)*2.)
      DO 5 I=1,NF
      SCSR=1.DO
      SCSI=0.DO
      DO 10 K=1,M
      ANG=-TPI*K*F*DT
      SCSR=SCSR+AR(K)*DCOS(ANG)-AI(K)*DSIN(ANG)
10    SCSI=SCSI+AR(K)*DSIN(ANG)+AI(K)*DCOS(ANG)
```

```
      EXTA(500+I)=PM*DT/(SCSR*SCSR+SCSI*SCSI)
 5    F=F+FINC
      RETURN
      END
      SUBROUTINE FGR(ISW)
      DOUBLE PRECISION XR(25),XI(25),FR(25),FI(25),
     + BR(25),BI(25),THETA(20),AR(10),AI(10),
     + GR(10,10),GI(10,10),PS(20),PGPR(10,10),
     + PGPI(10,10),PGMR(10,10),PGMI(10,10)
     + ,U,PM,TMPR,TMPI,TMQR,TMQI,T1R,T1I,
     + T2R,T2I,P1R,P1I,P2R,P2I,POWER,EXTA(820)
      COMMON /BLK0/N,M,U,PM,POWER/BLK1/XR,XI/BLK2/EXTA
      EQUIVALENCE (EXTA(1),GR(1,1)),(EXTA(101),GI(1,1)),
     + (EXTA(201),AR(1)),(EXTA(211),AI(1)),(EXTA(221),THETA(1)),
     + (EXTA(241),PS(1)),(EXTA(261),FR(1)),(EXTA(301),FI(1)),
     + (EXTA(341),BR(1)),(EXTA(381),BI(1)),
     + (EXTA(421),PGPR(1,1)),(EXTA(521),PGPI(1,1)),
     + (EXTA(621),PGMR(1,1)),(EXTA(721),PGMI(1,1))
      DO 310 J=1,M
      GR(J,J)=U*DSIN(THETA(J))*DCOS(THETA(M+J))
310   GI(J,J)=U*DSIN(THETA(J))*DSIN(THETA(M+J))
      DO 320 J=2,M
      J1=J-1
      DO 320 K=1,J1
      CALL CM(GR(J,J),GI(J,J),GR(J-1,J-K),-GI(J-1,J-K)
     + ,TMPR,TMPI)
320   CALL CAD(GR(J-1,K),GI(J-1,K),TMPR,TMPI,
     + GR(J,K),GI(J,K))
      IF(ISW.EQ.0)GO TO 377
      DO 379 J=1,M
      AR(J)=GR(M,J)
379   AI(J)=GI(M,J)
      RETURN
377   PM=0.D0
      NM=N-M
      DO 410 K=1,NM
      FR(K)=XR(K+M)
      FI(K)=XI(K+M)
      BR(K)=XR(K)
      BI(K)=XI(K)
      DO 420 I=1,M
      CALL CM(XR(K+M-I),XI(K+M-I),GR(M,I),GI(M,I)
     + ,TMPR,TMPI)
      CALL CADE(FR(K),FI(K),TMPR,TMPI)
      CALL CM(XR(K+I),XI(K+I),GR(M,I),-GI(M,I)
     + ,TMPR,TMPI)
420   CALL CADE(BR(K),BI(K),TMPR,TMPI)
410   PM=PM+FR(K)*FR(K)+FI(K)*FI(K)+BR(K)*BR(K)+
     + BI(K)*BI(K)
      PM=PM/(2.D0*DFLOAT(NM))
      DO 5 J=1,M
      DO 510 I=1,M
      PGPR(1,I)=0.D0
```

```
      PGPI(I,I)=0.D0
      PGMR(I,I)=0.D0
      PGMI(I,I)=0.D0
      IF(I.EQ.J)PGPR(I,I)=1.D0
 510  IF(I.EQ.J)PGMR(I,I)=1.D0
      DO 550 I=2,M
      I1=I-1
      DO 550 K=1,I1
      CALL CM(GR(I-1,I-K),-GI(I-1,I-K),PGPR(I,I),
     +PGPI(I,I),TMPR,TMPI)
      CALL CM(GR(I,I),GI(I,I),PGPR(I-1,I-K),-PGPI(I-1,I-K)
     +,TMQR,TMQI)
      PGPR(I,K)=PGPR(I-1,K)+TMPR+TMQR
      PGPI(I,K)=PGPI(I-1,K)+TMPI+TMQI
      CALL CM(GR(I-1,I-K),-GI(I-1,I-K),
     +PGMR(I,I),PGMI(I,I),TMPR,TMPI)
      CALL CM(GR(I,I),GI(I,I),
     +-PGMR(I-1,I-K),PGMI(I-1,I-K),TMQR,TMQI)
      PGMR(I,K)=PGMR(I-1,K)+TMPR+TMQR
      PGMI(I,K)=PGMI(I-1,K)+TMPI+TMQI
 550  CONTINUE
      TMPR=U*DCOS(THETA(J))*DCOS(THETA(J+M))
      TMPI=U*DCOS(THETA(J))*DSIN(THETA(J+M))
      TMQR=U*DSIN(THETA(J))*(-DSIN(THETA(J+M)))
      TMQI=U*DSIN(THETA(J))*DCOS(THETA(J+M))
      DO 25 I=1,M
      PGPR(1,I)=TMPR*PGPR(M,I)+TMPI*(-PGMI(M,I))
      PGPI(1,I)=TMPR*PGPI(M,I)+TMPI*PGMR(M,I)
      PGMR(1,I)=TMQR*PGPR(M,I)+TMQI*(-PGMI(M,I))
 25   PGMI(1,I)=TMQR*PGPI(M,I)+TMQI*PGMR(M,I)
      PS(J)=0.D0
      PS(J+M)=0.D0
      DO 20 K=1,NM
      T1R=0.D0
      T1I=0.D0
      T2R=0.D0
      T2I=0.D0
      P1R=0.D0
      P1I=0.D0
      P2R=0.D0
      P2I=0.D0
      DO 30 I=1,M
      CALL CM(XR(K+M-I),-XI(K+M-I),
     + PGPR(1,I),-PGPI(1,I),TMPR,TMPI)
      CALL CADE(T1R,T1I,TMPR,TMPI)
      CALL CM(XR(K+I),-XI(K+I),
     + PGPR(1,I),PGPI(1,I),TMPR,TMPI)
      CALL CADE(T2R,T2I,TMPR,TMPI)
      CALL CM(XR(K+M-I),-XI(K+M-I),
     + PGMR(1,I),-PGMI(1,I),TMPR,TMPI)
      CALL CADE(P1R,P1I,TMPR,TMPI)
      CALL CM(XR(K+I),-XI(K+I),PGMR(1,I),PGMI(1,I),TMPR,TMPI
 30   CALL CADE(P2R,P2I,TMPR,TMPI)
      CALL CM(FR(K),FI(K),T1R,T1I,TMPR,TMPI)
```

```
      CALL CM(BR(K),BI(K),T2R,T2I,TMQR,TMQI)
      PS(J)=PS(J)+TMPR+TMQR
      CALL CM(FR(K),FI(K),P1R,P1I,TMPR,TMPI)
      CALL CM(BR(K),BI(K),P2R,P2I,TMQR,TMQI)
20    PS(J+M)=PS(J+M)+TMPR+TMQR
      PS(J)=PS(J)/DFLOAT(NM)
5     PS(J+M)=PS(J+M)/DFLOAT(NM)
      RETURN
      END
      SUBROUTINE CM(XR,XI,YR,YI,ZR,ZI)
      DOUBLE PRECISION XR,XI,YR,YI,ZR,ZI
      ZR=XR*YR-XI*YI
      ZI=XR*YI+XI*YR
      RETURN
      END
      SUBROUTINE CAD(XR,XI,YR,YI,ZR,ZI)
      DOUBLE PRECISION XR,XI,YR,YI,ZR,ZI
      ZR=XR+YR
      ZI=XI+YI
      RETURN
      END
      SUBROUTINE CADE(XR,XI,YR,YI)
      DOUBLE PRECISION XR,XI,YR,YI
      XR=XR+YR
      XI=XI+YI
      RETURN
      END
      SUBROUTINE OPTM(LIMIT,LG)
      DOUBLE PRECISION H(85),X(20),GRATH(20)
     + ,F,OLDF,GRASUM,T,SQF,DY,HNRM,GNRM,EST,EPS
     + ,FY,ALFA,AMBDA,FX,DX,Z,DALFA,W,UER,EXTA(760),EXTS
      COMMON /BLK0/LX,N,UER,F,EXTS/BLK2/EXTA
      EQUIVALENCE (EXTA(221),X(1)),(EXTA(241),GRATH(1))
      LUN=6
      IF(LG.EQ.-1)LUN=5
      EPS=10.D0**(-16)
      WRITE(6,2222)
2222  FORMAT(1X,//,' OPTIMIZATION',//)
      WRITE(LUN,5)
5     FORMAT(' COUNT  RMS GRADIENT  RMS ERROR')
      CALL FGR(0)
      IER=0
      KOUNT=0
      N2=N+N
      N3=N2+N
      N31=N3+1
10    K=N31
      DO 40 J=1,N
      H(K)=1.D0
      NJ=N-J
      IF(NJ)50,50,20
20    DO 30 L=1,NJ
      KL=K+L
30    H(KL)=0.D0
```

```
40      K=KL+1
50      KOUNT=KOUNT+1
        OLDF=F
        GRASUM=0.D0
        DO 90 J=1,N
        GRASUM=GRASUM+GRATH(J)**2
        K=N+J
        H(K)=GRATH(J)
        K=K+N
        H(K)=X(J)
        K=J+N3
        T=0.D0
        DO 80 L=1,N
        T=T-GRATH(L)*H(K)
        IF(L-J)60,70,70
60      K=K+N-L
        GO TO 80
70      K=K+1
80      CONTINUE
90      H(J)=T
        GRASUM=DSQRT(GRASUM/DFLOAT(N))
        SQF=DSQRT(F)
        IF(MOD(KOUNT,1).EQ.0)WRITE(LUN,670)KOUNT,GRASUM,SQF
        DY=0.D0
        HNRM=0.D0
        GNRM=0.D0
        DO 100 J=1,N
        HNRM=HNRM+DABS(H(J))
        GNRM=GNRM+DABS(GRATH(J))
100     DY=DY+H(J)*GRATH(J)
        IF(DY)110,540,540
110     IF(HNRM/GNRM-EPS)540,540,120
120     FY=F
        ALFA=2.D0*(EST-F)/DY
        AMBDA=1.D0
        IF(ALFA)150,150,130
130     IF(ALFA-AMBDA)140,150,150
140     AMBDA=ALFA
150     ALFA=0.D0
160     FX=FY
        DX=DY
        DO 170 I=1,N
170     X(I)=X(I)+AMBDA*H(I)
        CALL FGR(0)
        FY=F
        DY=0.D0
        DO 180 I=1,N
180     DY=DY+GRATH(I)*H(I)
        IF(DY)190,390,220
190     IF(FY-FX)200,220,220
200     AMBDA=AMBDA+ALFA
        ALFA=AMBDA
        IF(HNRM*AMBDA-1.D10)160,160,210
```

```
210      IER=2
         RETURN
220      T=0.D0
230      IF(AMBDA)240,390,240
240      Z=3.D0*(FX-FY)/AMBDA+DX+DY
         ALFA=DMAX1(DABS(Z),DABS(DX),DABS(DY))
         DALFA=Z/ALFA
         DALFA=DALFA*DALFA-DX/ALFA*DY/ALFA
         IF(DALFA)540,250,250
250      W=ALFA*DSQRT(DALFA)
         ALFA=DY-DX+W+W
         IF(ALFA)260,270,260
260      ALFA=(DY-Z+W)/ALFA
         GO TO 280
270      ALFA=(Z+DY-W)/(Z+DX+Z+DY)
280      ALFA=ALFA*AMBDA
         DO 290 I=1,N
290      X(I)=X(I)+(T-ALFA)*H(I)
         CALL FGR(0)
         IF(F-FX)300,300,310
300      IF(F-FY)390,390,310
310      DALFA=0.D0
         DO 320 I=1,N
320      DALFA=DALFA+GRATH(I)*H(I)
         IF(DALFA)330,360,360
330      IF(F-FX)350,340,360
340      IF(DX-DALFA)350,390,350
350      FX=F
         DX=DALFA
         T=ALFA
         AMBDA=ALFA
         GO TO 230
360      IF(FY-F)380,370,380
370      IF(DY-DALFA)380,390,380
380      FY=F
         DY=DALFA
         AMBDA=AMBDA-ALFA
         GO TO 220
390      IF((OLDF-F)/OLDF-EPS*1.D-20)400,400,400
400      DO 410 J=1,N
         K=N+J
         H(K)=GRATH(J)-H(K)
         K=N+K
410      H(K)=X(J)-H(K)
         IER=0
         IF(KOUNT-N)450,420,420
420      T=0.D0
         Z=0.D0
         DO 430 J=1,N
         K=N+J
         W=H(K)
         K=K+N
         T=T+DABS(H(K))
```

```
 430      Z=Z+W*H(K)
          IF(HNRM-EPS)440,440,450
 440      IF(T-EPS)590,590,450
 450      IF(KOUNT-LIMIT)460,530,530
 460      ALFA=0.D0
          DO 500 J=1,N
          K=J+N3
          W=0.D0
          DO 490 L=1,N
          KL=N+L
          W=W+H(KL)*H(K)
          IF(L-J)470,480,480
 470      K=K+N-L
          GO TO 490
 480      K=K+1
 490      CONTINUE
          K=N+J
          ALFA=ALFA+W*H(K)
 500      H(J)=W
          IF(Z*ALFA)510,10,510
 510      K=N31
          DO 520 L=1,N
          KL=N2+L
          DO 520 J=L,N
          NJ=N2+J
          H(K)=H(K)+H(KL)*H(NJ)/Z-H(L)*H(J)/ALFA
 520      K=K+1
          GO TO 50
 530      IER=1
          RETURN
 540      DO 550 J=1,N
          K=N2+J
 550      X(J)=H(K)
          CALL FGR(0)
          IF(GNRM-EPS)580,580,560
 560      IF(IER.LT.-2)GO TO 590
          IER=IER-1
          GO TO 10
 580      IER=0
 585      FORMAT(' CONVERGENCE OCCURRED IN FMFP   ')
          WRITE(LUN,585)
 670      FORMAT(I6,2E20.12)
 590      RETURN
          END
          DOUBLE PRECISION FUNCTION DFLOAT(I)
          DFLOAT=DBLE(FLOAT(I))
          RETURN
          END
          SUBROUTINE MESG
          DOUBLE PRECISION GR(10,10),GI(10,10),AR(10),AI(10)
     +    ,EXTA(760),U,PM,POWER
          COMMON /BLK0/LX,LA,U,PM,POWER/BLK2/EXTA
          EQUIVALENCE (EXTA(1),GR(1,1)),(EXTA(101),GI(1,1))
     +    ,(EXTA(201),AR(1)),(EXTA(211),AI(1))
```

```
      WRITE(6,289)
289   FORMAT(1X,///,' LX,LA,U,PM,POWER',//)
      WRITE(6,251)LX,LA,U,PM,POWER
251   FORMAT(2X,2I10,3D16.7)
      WRITE(6,991)
991   FORMAT(1X,///,' THE DIAGONAL ELEMENTS OF G',//)
      WRITE(6,299)(GR(LJ,LJ),GI(LJ,LJ),LJ=1,LA)
299   FORMAT(1X,4(D12.5,3X,D12.5,2H ;),/)
      WRITE(6,378)
378   FORMAT(1X,///,' THE PEF COEFS. ',//)
      WRITE(6,299)(AR(LJ),AI(LJ),LJ=1,LA)
      RETURN
      END
```

APPENDIX C

Computer Program for Burg's Complex Signal Maximum Entropy Spectral Analysis

```
C   NAME: CMPBUG        DATE: 14-AUG-80
    COMMON INDEX
    REAL FNEW(32),GNEW(32),SF(513),S(513),YX(1000),CAL(1000)
    COMPLEX F(32),PEF(32),PER(32),GGG(20),GGGM(20),G(20),HALP(20)
    COMPLEX H(20)
    TYPE 5
5   FORMAT(1X,'***** I/P NUM,LG,SN,EXPAND,DT *****'/'0')
    ACCEPT 10,NUM,LG,SN,EXPAND,DT
10  FORMAT(2I5,3F12.5)
    TYPE 15
15  FORMAT(1X,'***** I/P ISTART,ISTOP,INC,NOUT,I2,LP *****'/'0')
    ACCEPT 20,ISTART,ISTOP,INC,NOUT,I2,LP
20  FORMAT(6I5)
    CALL BELL
    NI=NUM*2
    DEFINE FILE 2(2,NI,U,INDEX)
    CALL INPUT(F,NUM,SN,FNEW,GNEW)
    TPI=8.*ATAN(1.)
    CALL BELL
50  CONTINUE
    DEN=FLOAT(NOUT-1)*EXPAND
    KFUL=INT(DEN/EXPAND+1.)
    DO 60 KF=1,KFUL
    SF(KF)=FLOAT(KF-1)/DEN-.5
60  SF(KF)=TPI*SF(KF)
    DO 65 I=1,NUM
    FNEW(I)=REAL(F(I))
    GNEW(I)=-AIMAG(F(I))
65  CONTINUE
    DO 70 I=1,NUM
70  YX(I)=FLOAT(I)
    CALL NEWPAG
    CALL KBPLOT(YX,FNEW,NUM,50,1010,50,700)
    CALL FIG1(YX,CAL,348,3)
    CALL BELL
    TYPE 998
998 FORMAT(20X,'***** I/P CHANNEL I   WAVE  *****')
    ACCEPT 999,NX
999 FORMAT(I5)
    CALL NEWPAG
    CALL KBPLOT(YX,GNEW,NUM,50,1010,50,700)
    CALL FIG1(YX,CAL,348,3)
    CALL BELL
    TYPE 997
997 FORMAT(20X,'***** I/P CHANNEL II  WAVE *****')
    ACCEPT 999,NX
    P=0.
    DO 80 I=1,NUM
```

```
80  P=P+F(I)*CONJG(F(I))
    P=P/FLOAT(NUM)
    SAVE=P
    LAI=NUM-1
    DO 150 NN=1,LAI
    NNN=NN+1
    CALL BPEC(NUM,NN,PEF,PER,F,G,GGG,GGGM,H)
    P=(1.-G(NN)*CONJG(G(NN)))*P
    HELP=P
100 IF(NNN.GT.ISTOP)GO TO 150
    IF(NNN.LT.ISTART)GO TO 150
    IF(MOD((NNN-ISTART),INC).NE.0)GO TO 150
    CALL SPEC(NN,G,S,SF,KFUL,P,DT)
    CALL BELL
    CALL BELL
    ACCEPT 999,NX
    CALL NEWPAG
    CALL KBPLOT(YX,S,NOUT,50,1010,50,700)
    TYPE 120,NNN
120 FORMAT(30X,'***** NO. OF WEIGHT = ',I3,'*****')
    CALL FIG1(YX,CAL,I2,2)
    IF(LP.EQ.1)CALL PRIT(NOUT,DT,S,YX)
    IF(LG.LE.0)GO TO 140
    ACCEPT 999,NX
    CALL ALG(S,NOUT)
    CALL NEWPAG
    CALL KBPLOT(YX,S,NOUT,50,1010,50,700)
    TYPE 120,NNN
    CALL FIG1(YX,CAL,I2,1)
140 P=HELP
150 CONTINUE
    CALL BELL
    CALL EXIT
    END
    SUBROUTINE INPUT(X,N1,SN,Y,Z)
    COMMON INDEX
    COMPLEX X(1)
    REAL Y(N1),Z(N1)
    INDEX=1
    READ(2'INDEX)Y
    INDEX=2
    READ(2'INDEX)Z
    DO 5 KK=1,N1
5   Z(KK)=-Z(KK)
    DO 10 I=1,N1
    X(I)=CMPLX(Y(I),Z(I))+SN*CMPLX(GAUSS(0.,1.),GAUSS(0.,1.))
10  CONTINUE
    RETURN
    END
    FUNCTION GAUSS(XBAR,SIGMA)
    ROOT=1./SQRT(8.*ATAN(1.))
10  CONTINUE
    X=10.*(RAN(0,1)-.5)
    Y=ROOT*EXP( X*X/2.)
```

```
      YTRY=ROOT*RAN(0,1)
      IF(YTRY.GT.Y)GO TO 10
      GAUSS=X*SIGMA+XBAR
      RETURN
      END
      SUBROUTINE BPEC(NUM,NN,PEF,PER,F,G,GGG,GGGM,H)
      DIMENSION H(1),PEF(1),PER(1),F(1),G(1),GGG(1),GGGM(1)
      COMPLEX G,PEF,PER,H,GGG,GGGM,F,P,Q,SN,SD
      XERO=(0.,0.)
      N=NN-1
      IF(N.NE.0)GO TO 20
      DO 10 J=1,NUM
      PEF(J)=XERO
10    PER(J)=XERO
20    SN=XERO
      SD=XERO
      JJ=NUM-N-1
      DO 30 J=1,JJ
      Q=F(J+N+1)+PEF(J)
      P=F(J)+PER(J)
      SN=SN+CONJG(P)*Q
      SD=SD+P*CONJG(P)+Q*CONJG(Q)
30    CONTINUE
      GGG(NN)=-2.*SN/SD
      IF(N.EQ.0)GGGM(1)=GGG(1)
      IF(N.EQ.0)GO TO 60
      DO 40 J=1,N
      K=N-J+1
      H(J)=GGG(J)+GGG(NN)*CONJG(GGG(K))
40    CONTINUE
      DO 50 J=1,N
50    GGG(J)=H(J)
      GGGM(NN)=GGG(NN)
      JJ=JJ-1
60    DO 70 J=1,JJ
      PER(J)=PER(J)+CONJG(GGG(NN))*(PEF(J)+F(J+NN))
      PEF(J)=PEF(J+1)+GGG(NN)*(PER(J+1)+F(J+1))
70    CONTINUE
      DO 80 I=1,NN
80    G(I)=GGG(I)
      RETURN
      END
      SUBROUTINE SPEC(NN,G,S,SF,KFUL,P,DT)
      DIMENSION S(1),SF(1),G(1)
      COMPLEX G,SCS,ONE,GK
      ONE=(1.,0.)
      PQ=2.*P*DT
      DO 20 KF=1,KFUL
      CA=1.
      SA=0.
      SCS=ONE
      ADD=SF(KF)
      CT=COS(ADD)
      ST=SIN(ADD)
```

```
      DO 10 K=1,NN
      GK=G(K)
      TEMP=CT*CA-ST*SA
      SA=CT*SA+ST*CA
      CA=TEMP
      SCS=SCS+GK*CMPLX(CA,SA)
10    CONTINUE
      S(KF)=PQ/(SCS*CONJG(SCS))
20    CONTINUE
      RETURN
      END
      SUBROUTINE PRIT(NOUT,DT,S,YX)
      REAL S(1),YX(1)
      FREQ1=1./FLOAT(NOUT-1)/DT
      YX(1)=-0.50
      DO 10 I=2,NOUT
10    YX(I)=YX(I-1)+FREQ1
      DO 30 J=1,NOUT
      WRITE(6,20)J,YX(J),S(J)
20    FORMAT(10X,I3,5X,E15.7,5X,E15.7)
30    CONTINUE
      RETURN
      END
      SUBROUTINE FIG1(YX,CAL,I2,KG)
      REAL CAL(1),YX(1)
      CAL(1)=330.
      DO 100 I=1,960
      IF(I.EQ.1)GO TO 100
      I1=I-I/I2*I2
      IF(I1.EQ.0)GO TO 50
      CAL(I)=0.
      GO TO 100
50    CAL(I)=5.
100   CONTINUE
      DO 200 J=1,960
200   YX(J)=FLOAT(J)
      CALL KBPLOT(YX,CAL,960,50,1010,50,725)
      IF(KG.EQ.2)TYPE 205
205   FORMAT(5X,'LINEAR SCALE')
      IF(KG.EQ.1)TYPE 210
210   FORMAT(5X,'LOG. SCALE')
      IF(KG.EQ.3)TYPE 215
215   FORMAT(20X,'TIME  SCALE')
      CALL BELL
      RETURN
      END
      SUBROUTINE ALG(S,KFUL)
      REAL S(1)
      DO 10 I=1,KFUL
      IF(S(I).GT.0.)S(I)=ALOG10(S(I))
10    CONTINUE
      RETURN
      END
```

APPENDIX D

Computer Program for Complex Signal Multichannel Maximum Entropy Spectral Analysis

```
C+++++++++++++++++++++++++++++++++++++++++++++++++++++++
C
C       MULTICHANNEL MAXIMUM ENTROPY ROUTINE
C       BURG (GENERAL) P-CHANNEL, COMPLEX.
C       ARGUMENTS:
C          NC   MAX. FILTER LENGTH.
C          NP   # OF CHANNELS.
C          NSPE MIN. FILTER LENGTH(DESIRED SPECTRUM).
C          NDAT # OF DATA POINTS.
C          NPRS MIN. FILTER LENGTH(DESIRED PRINTOUT).
C
C       INPUT DATA MUST STORED IN FILE 'SIN.DAT'.
C
C       THIS PROGRAM MUST LINK WITH THE FOLLOWING SUBROUTINES:
C          DRAW        PROGRM TO DISPLAY THE DESIRED DATA
C                      ON THE TERMINAL.
C          VCVTCH
C          EIGCH       SUBROUTINES OF IMSL SOFTWARE PACKAGE.
C
C++++++++++++++++++++++++++++++++++++++++++++++++++++++++++++
      COMMON/TSER/YT,DT,NDAT,NP
      COMMON/DA/NC
      COMMON/SPECTC/NSPE
      COMMON/NPRINT/NPRS
      DIMENSION YT(4,300)
      DIMENSION UU(4),VV(4)
      COMPLEX YT
        TYPE 1110
1110    FORMAT(2X,'I/P NC,NP,NSPE,NDAT,NPRS,DT')
        ACCEPT *,NC,NP,NSPE,NDAT,NPRS,DT
      TYPE 4
    4 FORMAT(/)
      TYPE 15,NC,NP
   15 FORMAT(1X,27HMAX. NO. OF FILTER COEFF. =,1I5,2X,41HNO. OF COMPONE
     1NTS IN VECTOR TIME SERIES =,1I5,//)
      TYPE 7,NDAT,DT
    7 FORMAT(1X,24HORIGINAL TIME SERIES HAS,1I7,2X,28HDATA POINTS AT TI
     1ME INTERVAL,1F10.5,/)
        OPEN(UNIT=30,FILE='SIN.DAT')
      DO 2 K=1,NDAT
      READ (30,20)(UU(I),VV(I),I=1,NP)
   20 FORMAT(4F10.3)
      DO 3 I=1,NP
    3 YT(I,K)=CMPLX(UU(I),VV(I))
    2 CONTINUE
      CALL DETREN
      CALL CALFIL
        CALL EXIT
      END
```

```
C*************************************************************
      SUBROUTINE DETREN
      COMMON/TSER/YT,DT,NDAT,NP
      DIMENSION YT(4,300),X(300),Y(300)
      COMPLEX YT,ENUM,V10,V11,V12,V13,EMY,A
      ENDAT=FLOAT(NDAT)
      TYPE 1
    1 FORMAT(/)
      TYPE 2,NDAT,DT
    2 FORMAT(1X,34HNO. OF DATA POINTS ACTUALLY USED =,1I10,3X,38HWITH R
     1ESULTING TIME INTERVAL DELTA T =,1F17.10,/)
      TYPE 3,NP
    3 FORMAT(1X,36HTHIS COMPLEX VECTOR TIME SERIES HAS ,1I5,2X,10HCOMPO
     1NENTS,/)
        V1=(DT*DT)/12.
        V2=ENDAT+1.
        V2=V2*(ENDAT-1.)
        V2=V2*ENDAT
      ST2=V1*V2
      DO 4 K=1,NP
      TYPE 1
      TYPE 6,K
    6 FORMAT(1X,41HPRINTOUT BELOW PERTAINS TO COMPONENT NO. ,1I5,3X,14H
     1OF TIME SERIES,///)
      EMY=CMPLX(0.,0.)
      DO 5 I=1,NDAT
        X(I)=REAL(YT(K,I))
        Y(I)=AIMAG(YT(K,I))
    5 EMY=EMY+YT(K,I)
      EMY=EMY/ENDAT
      TYPE 10
   10 FORMAT(1X,50HREAL PARTS OF ORIGINAL TIME SERIES COMPONENT BELOW)
      TYPE 1
      TYPE 11,(X(I),I=1,NDAT)
   11 FORMAT(10F12.7)
      TYPE 1
      TYPE 12
   12 FORMAT(1X,55HIMAGINARY PARTS OF ORIGINAL TIME SERIES COMPONENT BE
     1LOW,/)
      TYPE 11,(Y(I),I=1,NDAT)
      TYPE 1
        EMYR=REAL(EMY)
        EMYI=AIMAG(EMY)
      TYPE 7,EMYR,EMYI
    7 FORMAT(1X,57HMEAN OF THIS COMPONENT OF RAW TIME SERIES HAS REAL P
     1ART =,1F15.7,2X,20HAND IMAGINARY PART =,1F15.7,/)
C  IF THE TWO CARDS ISKIP=1 AND IF(ISKIP.EQ.1)GO TO 30 ARE PLACED
C  BETWEEN THE 1,S BELOW,THE DATA WILL NOT BE DETRENDED. IF THEY ARE
C  PLACED BETWEEN THE 2,S,THE MEAN WILL BE SUBTRACTED.  IF BETWEEN
C  THE 3,S,THE DATA WILL BE GIVEN A FULL LINEAR DETREND.
C 1 1 1 1 1 1 1 1 1 1 1 1 1 1 NO DETREND 1 1 1 1 1 1 1 1 1 1 1 1 1 1 1 1
      ISKIP=1
      IF(ISKIP.EQ.1)GO TO 30
C 1 1 1 1 1 1 1 1 1 1 1 1 1 1 NO DETREND 1 1 1 1 1 1 1 1 1 1 1 1 1 1 1 1
```

```
C       TYPE 8
    8 FORMAT(5X,58HMEAN HAS BEEN SUBTRACTED FROM EACH VALUE OF THIS COM
     1PONENT,/)
      DO 9 I=1,NDAT
      YT(K,I)=YT(K,I)-EMY
      X(I)=REAL(YT(K,I))
    9 Y(I)=AIMAG(YT(K,I))
      TYPE 13
   13  FORMAT(1X,54HREAL PARTS OF MEAN-REDUCED TIME-SERIES COMPONENT BE
     1LOW,/)
      TYPE 11,(X(I),I=1,NDAT)
      TYPE 1
      TYPE 14
   14 FORMAT(1X,59HIMAGINARY PARTS OF MEAN-REDUCED TIME-SERIES COMPONEN
     1T BELOW,/)
      TYPE 11,(Y(I),I=1,NDAT)
      TYPE 1
C    2 2 2 2 2 2 2 2 2 2 2 SUBTRACT MEAN 2 2 2 2 2 2 2 2 2 2 2 2 2 2 2 2
C    2 2 2 2 2 2 2 2 2 2 2 SUBTRACT MEAN 2 2 2 2 2 2 2 2 2 2 2 2 2 2 2 2
        T=(-(ENDAT-1.)/2.)*DT
        A=CMPLX(0.,0.)
      DO 15 I=1,NDAT
      A=A+YT(K,I)*T
   15 T=T+DT
        A=A/ST2
        AR=REAL(A)
        AI=AIMAG(A)
      TYPE 16,AR,AI
   16 FORMAT(1X,50HREGRESSION COEFFICIENT FOR DETRENDING, REAL PART =,1
     1F12.6,2X,16HIMAGINARY PART =,1F12.6,/)
      TYPE 1
      T=(-(ENDAT-1.)/2.)*DT
      DO 17 I=1,NDAT
      YT(K,I)=YT(K,I)-A*T
        X(I)=REAL(YT(K,I))
        Y(I)=AIMAG(YT(K,I))
   17 T=T+DT
      TYPE 1
      TYPE 18
   18 FORMAT(1X,37HDETRENDED TIME SERIES COMPONENT BELOW,/)
      TYPE 19
   19 FORMAT(1X,51HREAL PARTS OF DETRENDED TIME SERIES COMPONENT BELOW,
     1/)
      TYPE 11,(X(I),I=1,NDAT)
      TYPE 20
   20 FORMAT(1X,56HIMAGINARY PARTS OF DETRENDED TIME-SERIES COMPONENT B
     1ELOW,/)
      TYPE 11,(Y(I),I=1,NDAT)
      TYPE 21
   21 FORMAT(1X,63HDETRENDED TIME-SERIES COMPONENT USED IN SUBSEQUENT C
     1ALCULATIONS,///)
C 3 3 3 3 3 3 3 3 3 3 3 3 3 3 3 3FULL DETREND 3 3 3 3 3 3 3 3 3 3 3 3 3
C 3 3 3 3 3 3 3 3 3 3 3 3 3 3 3 3FULL DETREND 3 3 3 3 3 3 3 3 3 3 3 3 3
   30 CONTINUE
```

```
      TYPE 1
    4 CONTINUE
      TYPE 1
      RETURN
      END
C*****************************************************************
      SUBROUTINE CALFIL
      COMMON/TSER/YT,DT,NDAT,NP
      COMMON/DA/NFF
      COMMON/SPECTC/NSPE
      COMMON/UPD/BT,ET,CB,CF,C,CP,E,B,G,SS,CN,CNP,BI,EI,BBB,I9,CNDAT,WM
     1,NN,N,N1,NPP,P,PP,NMI
      COMMON/POWPL/ELP,ELPP,FPE,DETP,DETPP
      COMMON/EIGIN/AEIG,WK,EINV
      COMPLEX YT,BT,ET,CB,CF,C,CP,E,B,G,SS,CN,CNP,BI,EI,BBB,V1,V2,V3,CN
     1DAT,WM,P,PP,AEIG,EINV
      DIMENSION YT(4, 300),BT(4, 300),ET(4, 300),CB(4,4,40),CF(4,4,40),
     1C(4,4),CP(4,4),E(4,4),B(4,4),G(4,4),SS(4,4),CN(4,4),CNP(4,4),BI(4,
     24),EI(4,4),BBB(16),V1(4,4),V2(4,4),V3(4,4),PR1(4,4),PR2(4,4),P(4,4
     3),PP(4,4),AEIG(4,4),WK(12),EINV(16,16)
      DIMENSION EL(4),ELP(40,2),ELPP(40,2),FPE(40),DETP(40),DETPP(40)
        NN=NDAT
        ENDAT=FLOAT(NDAT)
        CNDAT=CMPLX(ENDAT,0.)
        I9=1
        NM1=NDAT-1
        NPP=NP
      DO 1 I=1,NP
      DO 2 J=1,NP
      DO 3 K=2,40
      CB(I,J,K)=CMPLX(0.,0.)
    3 CF(I,J,K)=CMPLX(0.,0.)
      CB(I,J,1)=CMPLX(0.,0.)
    2 CF(I,J,1)=CMPLX(0.,0.)
      CB(I,I,1)=CMPLX(1.,0.)
      CF(I,I,1)=CMPLX(1.,0.)
    1 BBB(I)=CMPLX(0.,0.)
      DO 22 I=1,NP
      DO 23 J=1,NP
      SS(I,J)=CMPLX(0.,0.)
      DO 4 K=1,NDAT
    4 SS(I,J)=SS(I,J)+YT(I,K)*CONJG(YT(J,K))
   23 SS(I,J)=SS(I,J)/CNDAT
      DO 22 K=1,NDAT
      BT(I,K)=YT(I,K)
   22 ET(I,K)=YT(I,K)
      DO 24 I=1,NP
      DO 24 J=1,NP
      P(I,J)=SS(I,J)
   24 PP(I,J)=SS(I,J)
      DO 34 I=1,NP
      DO 34 J=1,NP
   34 AEIG(I,J)=SS(I,J)
      CALL VCVTCH(AEIG,NP,4,AEIG)
```

```
      CALL EIGCH(AEIG,NP,0,EL,V1,4,WK,IER)
        ELP(1,1)=EL(NP)
        ELPP(1,1)=EL(NP)
        ELP(1,2)=EL(1)
      ELPP(1,2)=EL(1)
      DETP(1)=1.
      DO 35 I=1,NP
   35 DETP(1)=DETP(1)*EL(I)
      DO 5 I=1,NP
      DO 5 J=1,NP
      PR1(I,J)=REAL(SS(I,J))
    5 PR2(I,J)=AIMAG(SS(I,J))
      TR=0.
      DO 6 I=1,NP
    6 TR=TR+PR1(I,I)
      FACT1=(ENDAT+1.)/(ENDAT-1.)
      FPE(1)=DETP(1)*(FACT1**NP)
      TYPE 8
    8 FORMAT(/)
      TYPE 7
    7 FORMAT(10X,29HOVERALL VARIANCE MATRIX BELOW,/)
      TYPE 9
    9 FORMAT(4(7X,4HREAL,10X,5HIMAG.,3X))
      TYPE 10,((PR1(I,J),PR2(I,J),J=1,NP),I=1,NP)
   10 FORMAT(8E15.6)
      TYPE 11,TR
   11 FORMAT(10X,7HTRACE =,1E20.10,//)
      DO 12 L=1,NFF
      I9=L
      CALL UPDAT
      CALL CPRINT
      IF(L.LT.NSPE)GO TO 33
      CALL SPECAL
   33 CONTINUE
   12 CONTINUE
      RETURN
      END
C***************************************************************
      SUBROUTINE SPECAL
      COMMON/TSER/YT,DT,NDAT,NP
      COMMON/DA/NFF
      COMMON/EIGIN/AEIG,WK,EINV
      COMMON/FFDATA/A(256),M2,N2,W(10)
      COMMON/UPD/BT,ET,CB,CF,C,CP,E,B,G,SS,CN,CNP,BI,EI,BBB,I9,CNDAT,WM
     1,NN,N,N1,NPP,P,PP,NMI
      COMPLEX YT,BT,ET,CB,CF,C,CP,E,B,G,SS,CN,CNP,BI,EI,BBB,V1I,V2I,V33
     1,V44,V55,V66,CNDAT,WM,P,PP,FFS,V1,EINV,AEIG
      COMPLEX A,W
      DIMENSION YT(4, 300),BT(4, 300),ET(4, 300),CB(4,4,40),CF(4,4,40),
     1C(4,4),CP(4,4),E(4,4),B(4,4),G(4,4),SS(4,4),CN(4,4),CNP(4,4),BI(4,
     24),EI(4,4),BBB(16),V1I(4,4),V2I(4,4),V33(4,4),V44(4,4),V55(4,4),V6
     36(4,4),P(4,4),PP(4,4),S(256,4),FREQ(256),FFS(4,4,256),EL(4),V1(4,4
     4),EINV(16,16),AEIG(4,4),WK(12)
      DIMENSION ELP(40,2),ELPP(40,2),FPE(40),DETP(40),DETPP(40)
```

```
      COMMON/POWPL/ELP,ELPP,FPE,DETP,DETPP
      COMMON/SIZEPL/XLOWER,XUPPER,YLOWER,YUPPER,KDIVX,KDIVY,KSCLX,KSCLY
      COMMON/LABPL/LABX(8),LABY(8),ID(8)
      COMMON/BDDPL/XBDD(2),YBDD(2)
      COMMON/CRTC/MXOP,MYOP,MFR
        N2=256
        M2=8
        N3=128
        ENDAT=FLOAT(NDAT)
        IFP1=I9+1
        N3P1=N3+1
        EN2=FLOAT(N2)
        DFF=1./(EN2*DT)
      DO 24 I=1,NP
      DO 24 J=1,NP
      DO 25 K=1,N2
   25 A(K)=CMPLX(0.,0.)
      DO 26 K=1,IFP1
   26 A(K)=CONJG(CF(J,I,K))
      CALL FFT
      DO 24 K=1,N2
   24 FFS(I,J,K)=A(K)
      DO 27 K=1,N2
      DO 28 I=1,NP
      DO 28 J=1,NP
   28 EINV(I,J)=FFS(I,J,K)
      CALL CMINV(EINV,NP,BBB)
      DO 29 I=1,NP
      DO 29 J=1,NP
      V33(I,J)=CMPLX(0.,0.)
      DO 29 KK=1,NP
   29 V33(I,J)=V33(I,J)+EINV(I,KK)*P(KK,J)
      DO 30 I=1,NP
      DO 30 J=1,NP
      V44(I,J)=CMPLX(0.,0.)
      DO 31 KK=1,NP
   31 V44(I,J)=V44(I,J)+V33(I,KK)*CONJG(EINV(J,KK))
   30 FFS(I,J,K)=V44(I,J)*DT
   27 CONTINUE
      DO 36 K=1,N3P1
      AKM1=FLOAT(K-1)
   36 FREQ(K)=DFF*AKM1
      DO 37 K=1,N3
        ENN=FLOAT(K-1-N3)
        N3PK=N3+K
   37 FREQ(N3PK)=DFF*ENN
      DO 60 K=1,N3
        N3PK=N3+K
        XX1=FREQ(K)
        XX2=FREQ(N3PK)
      DO 61 I=1,NP
      DO 61 J=1,NP
        V55(I,J)=FFS(I,J,K)
        V66(I,J)=FFS(I,J,N3PK)
```

```
      FFS(I,J,K)=V66(I,J)
   61 FFS(I,J,N3PK)=V55(I,J)
      FREQ(N3PK)=XX1
   60 FREQ(K)=XX2
      TYPE 38
   38 FORMAT(/)
      DO 40 K=1,N2
      DO 40 I=1,NP
   40 S(K,I)=REAL(FFS(I,I,K))
      DO 80 I=1,NP
      TYPE 39,I,I
   39 FORMAT(1X,I1,1H,,I1,21H-POWER SPECTRUM BELOW,/)
        ITT=I
        CALL DRAW(S,ITT,1)
      TYPE 41,(S(K,I),K=1,N2)
   41 FORMAT(10E13.5)
      TYPE 42
   80 CONTINUE
      TYPE 42
   42 FORMAT(/)
      IF(NP.EQ.1)GO TO 51
      NPM1=NP-1
      DO 82 I=1,NPM1
      IP1=I+1
      DO 82 J=IP1,NP
      DO 83 K=1,N2
        XX1=REAL(FFS(I,J,K))
        XX2=AIMAG(FFS(I,J,K))
      S(K,I)=CABS(FFS(I,J,K))
   83 S(K,J)=ATAN2(XX2,XX1)
      TYPE 42
      TYPE 44,I,J
   44 FORMAT(1X,14HABS. VALUE OF ,I2,1H,,I2,21H-POWER SPECTRUM BELOW,/)
        ITT=I
        CALL DRAW(S,ITT,1)
      TYPE 41,(S(K,I),K=1,N2)
      TYPE 42
      TYPE 45,I,J
   45 FORMAT(1X,13HARG.,RAD. FOR,I2,1H,,I2,21H-POWER SPECTRUM BELOW,/)
        ITT=J
        CALL DRAW(S,ITT,0)
      TYPE 41,(S(K,J),K=1,N2)
      TYPE 42
      DO 55 K=1,N2
        XX1=REAL(FFS(I,I,K))
        XX2=REAL(FFS(J,J,K))
   55 S(K,1)=((CABS(FFS(I,J,K)))**2)/(XX1*XX2)
      TYPE 56,I,J,I,I,J,J
   56 FORMAT(1X,25HSQUARED COHERENCE =(ABS(S,2I1,7H)**2)/S,2I1,2H*S,2I1
     1,6H BELOW,/)
        CALL DRAW(S,1,2)
      TYPE 41,(S(K,1),K=1,N2)
   82 CONTINUE
   51 CONTINUE
```

```
      TYPE 46
   46 FORMAT(5X,17HFREQUENCIES BELOW,/)
      TYPE 41,(FREQ(K),K=1,N2)
      TYPE 42
      DO 47 K=1,N2
      DO 48 I=1,NP
      DO 48 J=1,NP
   48 AEIG(I,J)=FFS(I,J,K)
      CALL VCVTCH(AEIG,NP,4,AEIG)
      CALL EIGCH(AEIG,NP,0,EL,V1,4,WK,IER)
      S(K,1)=EL(NP)
      S(K,2)=EL(1)
   47 CONTINUE
      TYPE 49
   49 FORMAT(1X,37HMAXIMAL EIGENVALUES OF SPECTRUM BELOW,/)
        CALL DRAW(S,1,1)
        TYPE 41,(S(K,1),K=1,N2)
      TYPE 42
      TYPE 50
   50 FORMAT(1X,37HMINIMAL EIGENVALUES OF SPECTRUM BELOW,/)
        CALL DRAW(S,2,1)
      TYPE 41,(S(K,2),K=1,N2)
      TYPE 38
      TYPE 70
   70 FORMAT(5X,35HDETERMINANTS OF FORWARD POWER BELOW,/)
      TYPE 41,(DETP(K),K=1,IFP1)
      TYPE 42
      DO 72 K=1,IFP1
   72 FREQ(K)=FLOAT(K-1)
      TYPE 42
      TYPE 172
  172 FORMAT(1X,25HAKAIKE FPE CRITERIA BELOW,/)
      TYPE 41,(FPE(I),I=1,IFP1)
      TYPE 42
      RETURN
      END
C*****************************************************************
      SUBROUTINE CPRINT
      COMMON/UPD/BT,ET,CB,CF,C,CP,E,B,G,SS,CN,CNP,BI,EI,BBB,I9,CNDAT,WM
     1,NN,N,N1,NPP,P,PP,NMI
      COMMON/POWPL/ELP,ELPP,FPE,DETP,DETPP
      COMMON/NPRINT/NPRS
      COMMON/EIGIN/AEIG,WK,EINV
      COMPLEX BT,ET,CB,CF,C,CP,E,B,G,SS,CN,CNP,BI,EI,BBB,V1,V2,V3,CNDAT
     1,WM,CNMI,CI,P,PP,AEIG,EINV
      DIMENSION BT(4, 300),ET(4, 300),CB(4,4,40),CF(4,4,40),C(4,4),CP(4
     1,4),E(4,4),B(4,4),G(4,4),SS(4,4),CN(4,4),CNP(4,4),BI(4,4),EI(4,4),
     2BBB(16),V1(4,4),V2(4,4),V3(4,4),P(4,4),PP(4,4),AEIG(4,4),EINV(16,1
     36)
      DIMENSION VBR(4,300),VBI(4,300),V4R(4,4),V4I(4,4),EL(4),WK(12)
      DIMENSION ELP(40,2),ELPP(40,2),FPE(40),DETP(40),DETPP(40)
        IFP1=I9+1
        NP=NPP
        NDAT=NMI+I9
```

```
      ENDAT=FLOAT(NDAT)
      EMEM=FLOAT(NP*I9)
    IF(I9.LT.NPRS)GO TO 80
    TYPE 1
  1 FORMAT(/)
    TYPE 2,I9,IFP1
  2 FORMAT(1X,18HRESULTS BELOW FOR ,1I5,3X,42HLAG(S),I.E. MATRIX OF B
   1LOCKWISE DIMENSION ,1I5,1X,1H.,/)
    TYPE 3
  3 FORMAT(10X,23HFORWARD RESIDUALS BELOW,/)
    DO 4 I=1,NP
    DO 4 J=1,NMI
    VBR(I,J)=REAL(ET(I,J))
  4 VBI(I,J)=AIMAG(ET(I,J))
    DO 7 I=1,NP
    TYPE 5,I
  5 FORMAT(5X,16HCOMPONENT NUMBER ,1I5,2X,5HBELOW,/)
    TYPE 6
  6 FORMAT(20X,16HREAL PARTS BELOW,/)
    TYPE 8,(VBR(I,J),J=1,NMI)
  8 FORMAT(10F13.8)
    TYPE 9
  9 FORMAT(/)
    TYPE 10
 10 FORMAT(15X,21HIMAGINARY PARTS BELOW,/)
    TYPE 8,(VBI(I,J),J=1,NMI)
    TYPE 1
  7 CONTINUE
    TYPE 50
 50 FORMAT(10X,24HBACKWARD RESIDUALS BELOW,/)
    DO 48 I=1,NP
    DO 48 J=1,NMI
    VBR(I,J)=REAL(BT(I,J))
 48 VBI(I,J)=AIMAG(BT(I,J))
    DO 49 I=1,NP
    TYPE 5,I
    TYPE 6
    TYPE 8,(VBR(I,J),J=1,NMI)
    TYPE 9
    TYPE 10
    TYPE 8,(VBI(I,J),J=1,NMI)
    TYPE 1
 49 CONTINUE
    TYPE 11
 11 FORMAT(10X,14HE-MATRIX BELOW,/)
    DO 12 I=1,NP
    DO 12 J=1,NP
    V4R(I,J)=REAL(E(I,J))
 12 V4I(I,J)=AIMAG(E(I,J))
    TYPE 6
    TYPE 13,((V4R(I,J),J=1,NP),I=1,NP)
 13 FORMAT(4E20.10)
    TYPE 10
    TYPE 13,((V4I(I,J),J=1,NP),I=1,NP)
```

```
      TYPE 9
      TR=0.
      DO 45 I=1,NP
   45 TR=TR+V4R(I,I)
      TYPE 47,TR
   47 FORMAT(10X,7HTRACE =,1E20.10,//)
      TYPE 14
   14 FORMAT(10X,14HG-MATRIX BELOW,/)
      DO 15 I=1,NP
      DO 15 J=1,NP
      V4R(I,J)=REAL(G(I,J))
   15 V4I(I,J)=AIMAG(G(I,J))
      TYPE 6
      TYPE 13,((V4R(I,J),J=1,NP),I=1,NP)
      TYPE 10
      TYPE 13,((V4I(I,J),J=1,NP),I=1,NP)
      TYPE 9
      TYPE 16
   16 FORMAT(10X,14HB-MATRIX BELOW,/)
      DO 17 I=1,NP
      DO 17 J=1,NP
      V4R(I,J)=REAL(B(I,J))
   17 V4I(I,J)=AIMAG(B(I,J))
      TYPE 6
      TYPE 13,((V4R(I,J),J=1,NP),I=1,NP)
      TYPE 10
      TYPE 13,((V4I(I,J),J=1,NP),I=1,NP)
      TYPE 9
      TR=0.
      DO 46 I=1,NP
   46 TR=TR+V4R(I,I)
      TYPE 47,TR
      TYPE 18
   18 FORMAT(1X,26HFORWARD POWER MATRIX BELOW,/)
      DO 19 I=1,NP
      DO 19 J=1,NP
      V4R(I,J)=REAL(P(I,J))
   19 V4I(I,J)=AIMAG(P(I,J))
      TYPE 6
      TYPE 13,((V4R(I,J),J=1,NP),I=1,NP)
      TYPE 10
      TYPE 13,((V4I(I,J),J=1,NP),I=1,NP)
      TYPE 9
      TYPE 20
   20 FORMAT(1X,27HBACKWARD POWER MATRIX BELOW,/)
      DO 21 I=1,NP
      DO 21 J=1,NP
      V4R(I,J)=REAL(PP(I,J))
   21 V4I(I,J)=AIMAG(PP(I,J))
      TYPE 6
      TYPE 13,((V4R(I,J),J=1,NP),I=1,NP)
      TYPE 10
      TYPE 13,((V4I(I,J),J=1,NP),I=1,NP)
      TYPE 9
```

```
      TYPE 22
   22 FORMAT(1X,43HFORWARD REFLECTION COEFFICIENT MATRIX BELOW,/)
      DO 23 I=1,NP
      DO 23 J=1,NP
      V4R(I,J)=REAL(CN(I,J))
   23 V4I(I,J)=AIMAG(CN(I,J))
      TYPE 6
      TYPE 13,((V4R(I,J),J=1,NP),I=1,NP)
      TYPE 10
      TYPE 13,((V4I(I,J),J=1,NP),I=1,NP)
      TYPE 9
      TYPE 24
   24 FORMAT(1X,44HBACKWARD REFLECTION COEFFICIENT MATRIX BELOW,/)
      DO 25 I=1,NP
      DO 25 J=1,NP
      V4R(I,J)=REAL(CNP(I,J))
   25 V4I(I,J)=AIMAG(CNP(I,J))
      TYPE 6
      TYPE 13,((V4R(I,J),J=1,NP),I=1,NP)
      TYPE 10
      TYPE 13,((V4I(I,J),J=1,NP),I=1,NP)
      TYPE 9
      DO 90 I=1,NP
      DO 90 J=1,NP
   90 AEIG(I,J)=E(I,J)
      CALL VCVTCH(AEIG,NP,4,AEIG)
      CALL EIGCH(AEIG,NP,0,EL,V1,4,WK,IER)
      TYPE 29
   29 FORMAT(1X,22HEIGENVALUES OF E BELOW,/)
      TYPE 13,(EL(I),I=1,NP)
      TYPE 9
      SUM=0.
      DO 60 I=1,NP
   60 SUM=SUM+EL(I)
      TYPE 38,SUM
   38 FORMAT(1X,13HSUM = TRACE =,1E20.10,//)
      TYPE 9
      DO 91 I=1,NP
      DO 91 J=1,NP
   91 AEIG(I,J)=B(I,J)
      CALL VCVTCH(AEIG,NP,4,AEIG)
      CALL EIGCH(AEIG,NP,0,EL,V1,4,WK,IER)
      SUM=0.
      DO 61 I=1,NP
   61 SUM=SUM+EL(I)
      TYPE 30
   30 FORMAT(1X,22HEIGENVALUES OF B BELOW,/)
      TYPE 13,(EL(I),I=1,NP)
      TYPE 9
      TYPE 38,SUM
   80 CONTINUE
      DO 92 I=1,NP
      DO 92 J=1,NP
   92 AEIG(I,J)=P(I,J)
```

```
      CALL VCVTCH(AEIG,NP,4,AEIG)
      CALL EIGCH(AEIG,NP,0,EL,V1,4,WK,IER)
      IF(EL(1).LT.0.)GO TO 94
      GO TO 95
   94 TYPE 96
   96 FORMAT(1X,48HMATRIX NOT POSITIVE DEFINITE--PROGRAM TERMINATED,/)
      TYPE 97,EL(1)
   97 FORMAT(1X,21HSMALLEST EIGENVALUE =,1E20.10)
      STOP
   95 CONTINUE
        SUM=0.
        DETP(IFP1)=1.
        ELP(IFP1,1)=EL(NP)
        ELP(IFP1,2)=EL(1)
      DO 62 I=1,NP
      DETP(IFP1)=DETP(IFP1)*EL(I)
   62 SUM=SUM+EL(I)
      FACT=((ENDAT+1.)+EMEM)/((ENDAT-1.)-EMEM)
      FPE(IFP1)=DETP(IFP1)*(FACT**NP)
      IF(I9.LT.NPRS)GO TO 81
      TYPE 33
   33 FORMAT(1X,22HEIGENVALUES OF P BELOW,/)
      TYPE 13,(EL(I),I=1,NP)
      TYPE 38,SUM
      TYPE 9
   81 CONTINUE
      DO 93 I=1,NP
      DO 93 J=1,NP
   93 AEIG(I,J)=PP(I,J)
      CALL VCVTCH(AEIG,NP,4,AEIG)
      CALL EIGCH(AEIG,NP,0,EL,V1,4,WK,IER)
      IF(EL(1).LT.0.)GO TO 104
      GO TO 105
  104 TYPE 96
      TYPE 97,EL(1)
      STOP
  105 CONTINUE
        SUM=0.
        DETPP(IFP1)=1.
        ELPP(IFP1,1)=EL(NP)
      ELPP(IFP1,2)=EL(1)
      DO 63 I=1,NP
      DETPP(IFP1)=DETPP(IFP1)*EL(I)
   63 SUM=SUM+EL(I)
      IF(I9.LT.NPRS)GO TO 82
      TYPE 36
   36 FORMAT(1X,28HEIGENVALUES OF P PRIME BELOW,/)
        TYPE 13,(EL(I),I=1,NP)
      TYPE 38,SUM
      TYPE 9
      TYPE 69
   69 FORMAT(1X,33HFORWARD FILTER COEFFICIENTS BELOW,/)
      DO 70 K=1,IFP1
      DO 71 I-1,NP
```

```
      DO 71 J=1,NP
      V4R(I,J)=REAL(CF(I,J,K))
   71 V4I(I,J)=AIMAG(CF(I,J,K))
      TYPE 6
      TYPE 13,((V4R(I,J),J=1,NP),I=1,NP)
      TYPE 10
      TYPE 13,((V4I(I,J),J=1,NP),I=1,NP)
   70 CONTINUE
      TYPE 73
   73 FORMAT(1X,34HBACKWARD FILTER COEFFICIENTS BELOW,/)
      DO 74 K=1,IFP1
      DO 75 I=1,NP
      DO 75 J=1,NP
      V4R(I,J)=REAL(CB(I,J,K))
   75 V4I(I,J)=AIMAG(CB(I,J,K))
      TYPE 6
      TYPE 13,((V4R(I,J),J=1,NP),I=1,NP)
      TYPE 10
      TYPE 13,((V4I(I,J),J=1,NP),I=1,NP)
   74 CONTINUE
      TYPE 9
   82 CONTINUE
      RETURN
      END
C***************************************************************
      SUBROUTINE UPDAT
      COMMON/UPD/BT,ET,CB,CF,C,CP,E,B,G,SS,CN,CNP,BI,EI,BBB,I9,CNDAT,WM
     1,NN,N,N1,NPP,P,PP,NMI
      COMMON/EIGIN/AEIG,WK,EINV
      COMPLEX U5,U6,AEIG,EINV
      COMPLEX BT,ET,CB,CF,C,CP,E,B,G,SS,CN,CNP,BI,EI,BBB,V1,V2,V3,CNDAT
     1,WM,CNMI,CI,P,PP,PII,PPI,Q1,Q2,Q1I,TI,U,AA,BB,CC,V4,V5
      DIMENSION BT(4, 300),ET(4, 300),CB(4,4,40),CF(4,4,40),C(4,4),CP(4
     1,4),E(4,4),B(4,4),G(4,4),SS(4,4),CN(4,4),CNP(4,4),BI(4,4),EI(4,4),
     2BBB(16),V1(4,4),V2(4,4),V3(4,4),P(4,4),PP(4,4),PII(4,4),PPI(4,4)
      DIMENSION V4(4,4),V5(4,4),V6(4,4),V7(4,4),Q1(4,4),Q2(4,4),Q1I(4,4
     1),TI(4,4),U(4,4),AA(4,4),BB(4,4),CC(4,4),AEIG(4,4),EINV(16,16)
      DIMENSION U5(4,4,40),U6(4,4,40),WK(12)
        AI=FLOAT(I9)
        CI=CMPLX(AI,0.)
        NDAT=NN
        NP=NPP
        NMI=NDAT-I9
        ENDAT=FLOAT(NDAT)
        CNDAT=CMPLX(ENDAT,0.)
        CNMI=CNDAT-CI
        WM=CMPLX(1.,0.)/CNMI
        IFP1=I9+1
        NMIM1=NMI-1
      DO 9 K=1,NMI
      KP1=K+1
      DO 9 I=1,NP
    9 ET(I,K)=ET(I,KP1)
      DO 1 I=1,NP
```

```
      DO 1 J=1,NP
        E(I,J)=CMPLX(0.,0.)
        B(I,J)=CMPLX(0.,0.)
        G(I,J)=CMPLX(0.,0.)
      DO 2 K=1,NMI
      E(I,J)=E(I,J)+ET(I,K)*CONJG(ET(J,K))
      B(I,J)=B(I,J)+BT(I,K)*CONJG(BT(J,K))
    2 G(I,J)=G(I,J)+BT(I,K)*CONJG(ET(J,K))
        E(I,J)=E(I,J)*WM
        B(I,J)=B(I,J)*WM
        G(I,J)=G(I,J)*WM
        PII(I,J)=P(I,J)
        PPI(I,J)=PP(I,J)
      BI(I,J)=B(I,J)
    1 EI(I,J)=E(I,J)
      DO 30 I=1,NP
      DO 30 J=1,NP
   30 EINV(I,J)=BI(I,J)
      CALL CMINV(EINV,NP,BBB)
      DO 31 I=1,NP
      DO 31 J=1,NP
   31 BI(I,J)=EINV(I,J)
      DO 32 I=1,NP
      DO 32 J=1,NP
   32 EINV(I,J)=EI(I,J)
      CALL CMINV(EINV,NP,BBB)
      DO 33 I=1,NP
      DO 33 J=1,NP
   33 EI(I,J)=EINV(I,J)
      DO 34 I=1,NP
      DO 34 J=1,NP
   34 EINV(I,J)=PII(I,J)
      CALL CMINV(EINV,NP,BBB)
      DO 35 I=1,NP
      DO 35 J=1,NP
   35 PII(I,J)=EINV(I,J)
      DO 36 I=1,NP
      DO 36 J=1,NP
   36 EINV(I,J)=PPI(I,J)
      CALL CMINV(EINV,NP,BBB)
      DO 37 I=1,NP
      DO 37 J=1,NP
   37 PPI(I,J)=EINV(I,J)
C  STEPS BETWEEN DASHED LINES DEFINE WEIGHT MATRICES Q1 AND Q2 AND
C   ALSO Q1 INVERSE
C   - - - - - - - - - - - - - - - - - - - - - - - - - - - - - -
      DO 10 I=1,NP
      DO 10 J=1,NP
      Q1(I,J)=PII(I,J)
      Q1I(I,J)=P(I,J)
   10 Q2(I,J)=PPI(I,J)
C   ABOVE STEPS GIVE INVERSE POWER WEIGHTING.
C  - - - - - - - - - - - - - - - - - - - - - - - - - - - - - - - - -
      DO 11 I=1,NP
```

```
      DO 11 J=1,NP
      V1(I,J)=CMPLX(0.,0.)
      DO 11 K=1,NP
   11 V1(I,J)=V1(I,J)+PP(I,K)*Q2(K,J)
      DO 12 I=1,NP
      DO 12 J=1,NP
      TI(I,J)=CMPLX(0.,0.)
      DO 12 K=1,NP
   12 TI(I,J)=TI(I,J)+V1(I,K)*PP(K,J)
      DO 38 I=1,NP
      DO 38 J=1,NP
   38 EINV(I,J)=TI(I,J)
      CALL CMINV(EINV,NP,BBB)
      DO 39 I=1,NP
      DO 39 J=1,NP
   39 TI(I,J)=EINV(I,J)
      DO 13 I=1,NP
      DO 13 J=1,NP
      V2(I,J)=CMPLX(0.,0.)
        V3(I,J)=CMPLX(0.,0.)
      DO 13 K=1,NP
      V2(I,J)=V2(I,J)+PII(I,K)*E(K,J)
   13 V3(I,J)=V3(I,J)+PII(I,K)*Q1I(K,J)
      DO 14 I=1,NP
      DO 14 J=1,NP
        U(I,J)=CMPLX(0.,0.)
        V5(I,J)=CMPLX(0.,0.)
      DO 14 K=1,NP
      U(I,J)=U(I,J)+V2(I,K)*PII(K,J)
   14 V5(I,J)=V5(I,J)+G(I,K)*V3(K,J)
      DO 15 I=1,NP
      DO 15 J=1,NP
      AA(I,J)=CMPLX(0.,0.)
        BB(I,J)=CMPLX(0.,0.)
        CC(I,J)=CMPLX(0.,0.)
      DO 15 K=1,NP
      AA(I,J)=AA(I,J)+TI(I,K)*B(K,J)
      BB(I,J)=BB(I,J)+U(I,K)*Q1I(K,J)
   15 CC(I,J)=(CC(I,J)-TI(I,K)*G(K,J))-PPI(I,K)*V5(K,J)
      CALL GETCN(AA,BB,CC,NP,CN)
      DO 25 I=1,NP
      DO 25 J=1,NP
      V1(I,J)=CMPLX(0.,0.)
      DO 25 K=1,NP
   25 V1(I,J)=V1(I,J)+CONJG(CN(K,I))*PP(K,J)
      DO 26 I=1,NP
      DO 26 J=1,NP
      CNP(I,J)=CMPLX(0.,0.)
      DO 26 K=1,NP
   26 CNP(I,J)=CNP(I,J)+PII(I,K)*V1(K,J)
      DO 27 I=1,NP
      DO 27 J=1,NP
      DO 27 K=1,IFP1
      U5(I,J,K)=CF(I,J,K)
```

```
   27 U6(I,J,K)=CB(I,J,K)
      DO 4 K=1,IFP1
      IP2MK=IFP1-K+1
      DO 4 I=1,NP
      DO 4 J=1,NP
      DO 4 KK=1,NP
      CF(I,J,K)=CF(I,J,K)+U6(I,KK,IP2MK)*CN(KK,J)
    4 CB(I,J,K)=CB(I,J,K)+U5(I,KK,IP2MK)*CNP(KK,J)
      DO 5 K=1,NMI
      DO 8 I=1,NP
      V3(I,1)=ET(I,K)
    8 V3(I,2)=BT(I,K)
      DO 5 I=1,NP
      DO 5 KK=1,NP
      ET(I,K)=ET(I,K)+CONJG(CN(KK,I))*V3(KK,2)
    5 BT(I,K)=BT(I,K)+CONJG(CNP(KK,I))*V3(KK,1)
      DO 6 I=1,NP
      DO 6 J=1,NP
      V1(I,J)=CMPLX(0.,0.)
        V2(I,J)=CMPLX(0.,0.)
      DO 6 K=1,NP
      V1(I,J)=V1(I,J)+PP(I,K)*CN(K,J)
    6 V2(I,J)=V2(I,J)+P(I,K)*CNP(K,J)
      DO 7 I=1,NP
      DO 7 J=1,NP
      DO 7 K=1,NP
      P(I,J)=P(I,J)-CONJG(CN(K,I))*V1(K,J)
    7 PP(I,J)=PP(I,J)-CONJG(CNP(K,I))*V2(K,J)
      RETURN
      END
C-----------------------------------------------------------------
      SUBROUTINE GETCN(A,B,C,N,CN)
      DIMENSION A(4,4),B(4,4),C(4,4),CN(4,4),BB(16),AB(16,16)
      COMPLEX A,B,C,BB,CN,AB
C  THIS SUBROUTINE SOLVES THE MATRIX EQUATION A*(CN)+(CN)*B=C FOR THE
C   REFLECTION COEFFICIENT,CN.   THE METHOD IS THAT OF BELLMAN,MATRIX
C   ANALYSIS,OLD EDITION,PAGE 231.
      N2=N*N
      DO 1 I=1,N
      DO 1 J=1,N
      I1=(I-1)*N+J
    1 BB(I1)=C(I,J)
      CALL KRONPR(A,B,AB,N)
      CALL CMINV(AB,N2,BB)
      DO 2 I=1,N
      DO 2 J=1,N
      I1=(I-1)*N+J
    2 CN(I,J)=BB(I1)
      RETURN
      END
C-----------------------------------------------------------------
      SUBROUTINE KRONPR(A,B,AB,N)
      DIMENSION A(4,4),B(4,4),AB(16,16)
      COMPLEX A,B,AB
```

```
C  THIS SUBROUTINE FORMS THE(N**2)X(N**2) MATRIX AB WHICH HAS THE
C  FORM AB=(AXI)+(IXBT) WHERE THE CROSSES DENOTE KRONECKER PRODUCTS.
C  SEE BELLMAN,MATRIX ANALYSIS,OLD EDITION,P. 231.
      N2=N*N
      DO 1 I=1,N2
      DO 1 J=1,N2
    1 AB(I,J)=CMPLX(0.,0.)
      DO 2 I=1,N
      DO 2 J=1,N
      DO 2 K=1,N
      I1=(I-1)*N+K
        I2=(J-1)*N+K
    2 AB(I1,I2)=A(I,J)
      DO 3 I=1,N
      DO 3 J=1,N
      DO 3 K=1,N
      I1=(I-1)*N+J
        I2=(I-1)*N+K
    3 AB(I1,I2)=AB(I1,I2)+B(K,J)
      RETURN
      END
C-------------------------------------------------------------------
      SUBROUTINE CMINV(A,N,B)
      DIMENSION IPIVOT(16),A(16,16),B(16),INDEX(16,2),PIVOT(16)
      COMPLEX AMAX,A,SWAP,B,PIVOT,T
      DO 20 J=1,N
   20 IPIVOT(J)=0
      DO 550 I=1,N
      AMAX=CMPLX(0.,0.)
      DO 101 J=1,N
      IF(IPIVOT(J).EQ.1)GO TO 101
      DO 100 K=1,N
      IF(IPIVOT(K)-1)80,100,740
   80 A11=CABS(AMAX)
      A12=CABS(A(J,K))
      IF(A11.LT.A12)85,100
   85 IROW=J
      ICOLUM=K
      AMAX=A(J,K)
  100 CONTINUE
  101 CONTINUE
      IPIVOT(ICOLUM)=IPIVOT(ICOLUM)+1
      IF(IROW.EQ.ICOLUM)GO TO 260
      DO 200 L=1,N
      SWAP=A(IROW,L)
      A(IROW,L)=A(ICOLUM,L)
  200 A(ICOLUM,L)=SWAP
      SWAP=B(IROW)
      B(IROW)=B(ICOLUM)
      B(ICOLUM)=SWAP
  260 INDEX(I,1)=IROW
      INDEX(I,2)=ICOLUM
      PIVOT(I)=A(ICOLUM,ICOLUM)
```

```
      A(ICOLUM,ICOLUM)=CMPLX(1.,0.)
      DO 350 L=1,N
  350 A(ICOLUM,L)=A(ICOLUM,L)/PIVOT(I)
      B(ICOLUM)=B(ICOLUM)/PIVOT(I)
      DO 550 L1=1,N
      IF(L1.EQ.ICOLUM)GO TO 550
  400 T=A(L1,ICOLUM)
      A(L1,ICOLUM)=CMPLX(0.,0.)
      DO 450 L=1,N
  450 A(L1,L)=A(L1,L)-A(ICOLUM,L)*T
      B(L1)=B(L1)-B(ICOLUM)*T
  550 CONTINUE
      DO 710 I=1,N
      L=N+1-I
      IF(INDEX(L,1).EQ.INDEX(L,2))GO TO 710
      JROW=INDEX(L,1)
      JCOLUM=INDEX(L,2)
      DO 705 K=1,N
      SWAP=A(K,JROW)
      A(K,JROW)=A(K,JCOLUM)
      A(K,JCOLUM)=SWAP
  705 CONTINUE
  710 CONTINUE
  740 RETURN
      END
C-------------------------------------------------------------------
      SUBROUTINE FFT
      COMMON/FFDATA/A(256),M,N,W(10)
      COMPLEX A,U,W,T
      COMMON/LASTN/LN
      LN=0
      IF(LN.NE.N) CALL INIFFT
      CALL REVBIT
      LE = 1
      DO 90 L = 1,M
         LE1 = LE
         LE = 2*LE
         U = CMPLX(1.,0.)
         DO 80 J = 1,LE1
            DO 70 I = J,N,LE
               IP = I+LE1
               T = A(IP)*U
               A(IP) = A(I) - T
               A(I)  = A(I) + T
   70       CONTINUE
            U = U*W(L)
   80    CONTINUE
   90 CONTINUE
      RETURN
      END
```

```
C-----------------------------------------------------------------
      SUBROUTINE INIFFT
      COMMON/FFDATA/A(256),M,N,W(10)
      COMMON/LASTN/LN
      COMPLEX A,U,W,T
      PI=3.1415926536
      LN = N
      EL = 1.
      DO 110 I = 1,M
         X = COS(PI/EL)
         Y = -SIN(PI/EL)
         EL = 2.*EL
         W(I) = CMPLX(X,Y)
  110 CONTINUE
      RETURN
      END
C-----------------------------------------------------------------
      SUBROUTINE REVBIT
^L      COMMON/FFDATA/A(256),M,N,W(10)
      COMPLEX A,W,T
      NV2 = N/2
      NM1 = N-1
      J = 1
      DO 3 I=1,NM1
^L          IF(I.GE.J) GO TO 1
             T = A(J)
             A(J) = A(I)
^L              A(I) = T
^L 1        CONTINUE
         K = NV2
 2       CONTINUE
            IF(K.GE.J) GO TO 25
            J = J - K
^L              K = K/2
^L          GO TO 2
 25      CONTINUE
         J = J + K
 3    CONTINUE
      RETURN
      END
```

APPENDIX E

Computer Program for Real Signal Multichannel Maximum Entropy Spectral Analysis

```
C          MUL13
           REAL SS(9,100)
           REAL ET(3,100),BT(3,100),UU(3)
           COMMON /BLK0/ET,BT
           COMMON /BLK1/SS
           DEFINE FILE 1(9,200,U,LINE2)
           DEFINE FILE 3(100,6,U,LINE)
           TYPE 10
10         FORMAT(////10X,'I/P N,NSPE,NFREQ',//)
           ACCEPT 20,N,NSPE,NFREQ
20         FORMAT(4I7)
           NDAT=100
           NP=3
           NOUT=100
           DT=1.
           DO 2 K=1,NDAT
           LINE=K
           READ(3'LINE)UU
           DO 3 I=1,NP
           ET(I,K)=UU(I)
3          BT(I,K)=UU(I)
2          CONTINUE
           CALL RMIN
           CALL STRAND(NOUT,NP,NDAT,N,DT,NFREQ,NSPE)
           DO 40 K=1,9
           LINE2=K
           WRITE(1'LINE2)(SS(K,J),J=1,100)
40         CONTINUE
           CALL EXIT
           END
           SUBROUTINE RMIN
           REAL ET(3,100),BT(3,100)
           COMMON /BLK0/ET,BT
           DO 50 I=1,3
           RM=0.
           DO 30 J=1,100
30         RM=RM+ET(I,J)
           RM=RM/100.
           DO 40 K=1,100
           ET(I,K)=ET(I,K)-RM
40         BT(I,K)=BT(I,K)-RM
50         CONTINUE
           RETURN
           END
```

```
      SUBROUTINE STRAND(NOUT,NP,NDAT,N,DT,NFREQ,NSPE)
      DIMENSION BT(3,100),ET(3,100),CF(3,3,25),CB(3,3,25),FPE(25)
     1,E(3,3),B(3,3),G(3,3),P(3,3),PP(3,3),Q1(3,3),V1(3,3),V3(3,3)
     2,CN(3,3),CNP(3,3),U5(3,3,25),U6(3,3,25),Q2(3,3),TI(3,3),AA(3,3
     3,BB(3,3),CC(3,3),SS(9,100)
      COMMON /BLKO/ET,BT
      COMMON /BLK1/SS
      NP1=N+1
      DO 11  I=1,NP
      DO 11  J=1,NP
      DO 11  K=1,NP1
      CB(I,J,K)=0.
   11 CF(I,J,K)=0.
      DO 31  I=1,NP
      CB(I,I,1)=1.
   31 CF(I,I,1)=1.
      DO 22  I=1,NP
      DO 22  J=1,NP
      P(I,J)=0.
      DO 21  K=1,NDAT
   21 P(I,J)=P(I,J)+ET(I,K)*ET(J,K)
      P(I,J)=P(I,J)/FLOAT(NDAT)
   22 PP(I,J)=P(I,J)
      CALL AKAIK(P,NP,NDAT,0.,FPE,1)
      DO 12 L=2,N
      LM1=L-1
      NDATM9=NDAT-LM1
      DO 9 K=1,NDATM9
      KP1=K+1
      DO 9 I=1,NP
    9 ET(I,K)=ET(I,KP1)
      DO 1  I=1,NP
      DO 1  J=1,NP
      E(I,J)=0.
      B(I,J)=0.
      G(I,J)=0.
      DO 2 K=1,NDATM9
      E(I,J)=E(I,J)+ET(I,K)*ET(J,K)
      B(I,J)=B(I,J)+BT(I,K)*BT(J,K)
    2 G(I,J)=G(I,J)+BT(I,K)*ET(J,K)
      E(I,J)=E(I,J)/FLOAT(NDATM9)
      B(I,J)=B(I,J)/FLOAT(NDATM9)
      G(I,J)=-2.*G(I,J)/FLOAT(NDATM9)
      Q1(I,J)=P(I,J)
    1 Q2(I,J)=PP(I,J)
      CALL INVERS(Q1,NP)
      CALL INVERS(Q2,NP)
      CALL ACCUM(NP,AA,Q2,B)
      CALL ACCUM(NP,BB,Q1,E)
      CALL ACCUM(NP,CC,Q2,G)
      CALL GETCNN(AA,BB,CC,NP,CN)
      DO 25 I=1,NP
      DO 25 J=1,NP
      V1(I,J)=0.
```

```
      DO 25 K=1,NP
25    V1(I,J)=V1(I,J)+CN(K,I)*PP(K,J)
      CALL ACCUM(NP,CNP,Q1,V1)
      DO 27 I=1,NP
      DO 27 J=1,NP
      DO 27 K=1,L
      U5(I,J,K)=CF(I,J,K)
27    U6(I,J,K)=CB(I,J,K)
      DO 4 K=1,L
      INDEX=L-K+1
      DO 4 I=1,NP
      DO 4 J=1,NP
      DO 4 KK=1,NP
      CF(I,J,K)=CF(I,J,K)+U6(I,KK,INDEX)*CN(KK,J)
      CB(I,J,K)=CB(I,J,K)+U5(I,KK,INDEX)*CNP(KK,J)
4     CONTINUE
      DO 5 K=1,NDATM9
      DO 8 I=1,NP
      V3(I,1)=ET(I,K)
8     V3(I,2)=BT(I,K)
      DO 5 I=1,NP
      DO 5 KK=1,NP
      ET(I,K)=ET(I,K)+CN(KK,I)*V3(KK,2)
5     BT(I,K)=BT(I,K)+CNP(KK,I)*V3(KK,1)
      CALL ACCUM(NP,V1,PP,CN)
      CALL ACCUM(NP,V3,P,CNP)
      DO 7 I=1,NP
      DO 7 J=1,NP
      DO 7 K=1,NP
      P(I,J)=P(I,J)-CN(K,I)*V1(K,J)
7     PP(I,J)=PP(I,J)-CNP(K,I)*V3(K,J)
      S=FLOAT(NP*LM1)
      CALL AKAIK(P,NP,NDAT,S,FPE,L)
      IF(LM1.LT.NSPE)GO TO 12
      TYPE 9999
9999  FORMAT(/5X,'1'/)
      CALL SPECAL(NOUT,LM1,NDAT,NP,CF,P,FPE,DT,NFREQ)
12    CONTINUE
      RETURN
      END
      SUBROUTINE SPECAL(NOUT,I9,NDAT,NP,CF,P,FPE,DT,NFREQ)
      REAL CF(3,3,25),P(3,3),FPE(25),TI(3,3),E(3,3),B(3,3),G(3,3)
     1,V1(3,3),V3(3,3),Q1(3,3),Q2(3,3)
      REAL SS(9,100)
      COMMON /BLK1/SS
      I9P1=I9+1
      ANGLE=3.1415926/FLOAT(NFREQ)
      FREQ1=1./(2.*FLOAT(NFREQ))
      DEGREE=360./(2.*3.1415926)
      C=1.
      S=0.
      C1=COS(ANGLE)
      S1=SIN(ANGLE)
      DO 99 KK=1,NOUT
```

```
      FREQ=FLOAT(KK)*FREQ1
      C2=C1*C-S1*S
      S=C1*S+S1*C
      C=C2
      CN1=1.
      SN1=0.
      DO 90 I=1,NP
      DO 90 J=1,NP
      E(I,J)=0.
90    B(I,J)=0.
      DO 91 L=1,ISP1
      DO 92 I=1,NP
      DO 92 J=1,NP
      E(I,J)=E(I,J)+CN1*CF(J,I,L)
92    B(I,J)=B(I,J)+SN1*CF(J,I,L)
      CN2=CN1*C-SN1*S
      SN1=CN1*S+SN1*C
      CN1=CN2
91    CONTINUE
      DO 93 I=1,NP
      DO 93 J=1,NP
93    V1(I,J)=E(I,J)
      CALL INVERS(V1,NP)
      CALL ACCUM(NP,V3,V1,B)
      CALL ACCUM(NP,G,B,V3)
      DO 95 I=1,NP
      DO 95 J=1,NP
95    Q1(I,J)=E(I,J)+G(I,J)
      CALL INVERS(Q1,NP)
      CALL ACCUM(NP,V3,B,V1)
      CALL ACCUM(NP,Q2,Q1,V3)
      DO 96 I=1,NP
      DO 96 J=1,NP
96    Q2(I,J)=-Q2(I,J)
      CALL ACCUM(NP,V1,Q1,P)
      CALL ACCUM(NP,V3,Q2,P)
      DO 30 I=1,NP
      DO 30 J=1,NP
      B(I,J)=0.
      G(I,J)=0.
      DO 30 K=1,NP
      B(I,J)=B(I,J)+V1(I,K)*Q1(J,K)+V3(I,K)*Q2(J,K)
30    G(I,J)=G(I,J)+V3(I,K)*Q1(J,K)-V1(I,K)*Q2(J,K)
      DO 31 I=1,NP
      DO 31 J=1,NP
      B(I,J)=2.*DT*B(I,J)
31    G(I,J)=2.*DT*G(I,J)
      CALL PHCO(G,B,NP,KK,FREQ)
99    CONTINUE
      TYPE 9999
9999  FORMAT(/5X,'2'/)
      ACCEPT 9998,NX
9998  FORMAT(I5)
      RETURN
      END
```

```
      SUBROUTINE PHCO(G,B,NP,KK,FREQ)
      REAL COH(3),PHAS(3)
      REAL G(3,3),B(3,3),SS(9,100)
      COMMON /BLK1/SS
      DEG=360./(2.*3.14159265)
      PHAS(1)=B(1,2)
      PHAS(2)=B(1,3)
      PHAS(3)=B(2,3)
      COH(1)=G(1,2)
      COH(2)=G(1,3)
      COH(3)=G(2,3)
      DO 20 J=1,NP
20    SS(J,KK)=B(J,J)
      DO 30 J=1,3
      SS(NP+J,KK)=PHAS(J)
30    SS(NP+3+J,KK)=COH(J)
      RETURN
      END
      SUBROUTINE GETCNN(A,B,C,N,CN)
      REAL A(3,3),B(3,3),C(3,3),CN(3,3),BB(9),AB(9,9)
      N2=N*N
      DO 1 I=1,N
      DO 1 J=1,N
      I1=(I-1)*N+J
1     BB(I1)=C(I,J)
      CALL KRONPR(A,B,AB,N)
      CALL CMINV(AB,N2,BB)
      DO 2 I=1,N
      DO 2 J=1,N
      I1=(I-1)*N+J
2     CN(I,J)=BB(I1)
      RETURN
      END
      SUBROUTINE CMINV(A,N,B)
      DIMENSION IPIVOT(9),A(9,9),B(9),INDEX(9,2),PIVOT(9)
      DO 20 J=1,N
20    IPIVOT(J)=0
      DO 550 I=1,N
      AMAX=0.
      DO 101 J=1,N
      IF(IPIVOT(J).EQ.1)GO TO 101
      DO 100 K=1,N
      IF(IPIVOT(K)-1)80,100,740
80    A11=ABS(AMAX)
      A12=ABS(A(J,K))
      IF(A11.GE.A12)GO TO 100
85    IROW=J
      ICOLUM=K
      AMAX=A(J,K)
100   CONTINUE
101   CONTINUE
      IPIVOT(ICOLUM)=IPIVOT(ICOLUM)+1
      IF(IROW.EQ.ICOLUM)GO TO 260
      DO 200 L=1,N
      SWAP=A(IROW,L)
```

```
      A(IROW,L)=A(ICOLUM,L)
200   A(ICOLUM,L)=SWAP
      SWAP=B(IROW)
      B(IROW)=B(ICOLUM)
      B(ICOLUM)=SWAP
260   INDEX(I,1)=IROW
      INDEX(I,2)=ICOLUM
      PIVOT(I)=A(ICOLUM,ICOLUM)
      A(ICOLUM,ICOLUM)=1.
      DO 350 L=1,N
350   A(ICOLUM,L)=A(ICOLUM,L)/PIVOT(I)
      B(ICOLUM)=B(ICOLUM)/PIVOT(I)
      DO 550 L1=1,N
      IF(L1.EQ.ICOLUM)GO TO 550
400   T=A(L1,ICOLUM)
      A(L1,ICOLUM)=0.
      DO 450 L=1,N
450   A(L1,L)=A(L1,L)-A(ICOLUM,L)*T
      B(L1)=B(L1)-B(ICOLUM)*T
550   CONTINUE
      DO 710 I=1,N
      L=N+1-I
      IF(INDEX(L,1).EQ.INDEX(L,2))GO TO 710
      JROW=INDEX(L,1)
      JCOLUM=INDEX(L,2)
      DO 705 K=1,N
      SWAP=A(K,JROW)
      A(K,JROW)=A(K,JCOLUM)
      A(K,JCOLUM)=SWAP
705   CONTINUE
710   CONTINUE
740   CONTINUE
      RETURN
      END
      SUBROUTINE INVERS(AINV,NP)
      DIMENSION AINV(3,3),EINV(9,9),BBB(9)
      DO 30 I=1,NP
      BBB(I)=0.
      DO 30 J=1,NP
30    EINV(I,J)=AINV(I,J)
      CALL CMINV(EINV,NP,BBB)
      DO 31 I=1,NP
      DO 31 J=1,NP
31    AINV(I,J)=EINV(I,J)
      RETURN
      END
      SUBROUTINE ACCUM(NP,A,B,C)
      DIMENSION A(3,3),B(3,3),C(3,3)
      DO 11 I=1,NP
      DO 11 J=1,NP
      A(I,J)=0.
      DO 11 K=1,NP
11    A(I,J)=A(I,J)+B(I,K)*C(K,J)
      RETURN
      END
```

```
      SUBROUTINE KRONPR(A,B,AB,N)
      DIMENSION A(3,3),B(3,3),AB(9,9)
      N2=N*N
      DO 1 I=1,N2
      DO 1 J=1,N2
1     AB(I,J)=0.
      DO 2 I=1,N
      DO 2 J=1,N
      DO 2 K=1,N
      I1=(I-1)*N+K
      I2=(J-1)*N+K
2     AB(I1,I2)=A(I,J)
      DO 3 I=1,N
      DO 3 J=1,N
      DO 3 K=1,N
      I1=(I-1)*N+J
      I2=(I-1)*N+K
3     AB(I1,I2)=AB(I1,I2)+B(K,J)
      RETURN
      END
      FUNCTION DETERM(ARRAY,NORDER)
      DIMENSION ARRAY(3,3)
      DETERM=1.
      DO 50 K=1,NORDER
      IF(ARRAY(K,K).NE.0.)GO TO 41
      DO 23 J=K,NORDER
      IF(ARRAY(K,J).NE.0.)GO TO 31
23    CONTINUE
      DETERM=0.
      RETURN
31    DO 34 I=K,NORDER
      SAVE=ARRAY(I,J)
      ARRAY(I,J)=ARRAY(I,K)
34    ARRAY(I,K)=SAVE
      DETERM=-DETERM
41    DETERM=DETERM*ARRAY(K,K)
      IF(K.GE.NORDER)GO TO 50
      K1=K+1
      DO 46 I=K1,NORDER
      DO 46 J=K1,NORDER
46    ARRAY(I,J)=ARRAY(I,J)-ARRAY(I,K)*ARRAY(K,J)/ARRAY(K,K)
50    CONTINUE
      RETURN
      END
      SUBROUTINE AKAIK(P,NP,NDAT,S,FPE,INDEX)
      DIMENSION P(3,3),FPE(25),DUM(3,3)
      DO 1 I=1,NP
      DO 1 J=1,NP
1     DUM(I,J)=P(I,J)
      DETP=DETERM(DUM,NP)
      FACT1=FLOAT(NDAT)+1.+S
      FACT2=FLOAT(NDAT)-1.-S
      FACT=FACT1/FACT2
      FPE(INDEX)=DETP*(FACT**NP)
      RETURN
      END
```

APPENDIX F

Computer Program for Two-Dimensional Maximum Entropy Spectral Analysis

```
C+++++++++++++++++++++++++++++++++++++++++++++++++++++
C
C       THIS PROGRAM IS TO CALCULATE THE AUTOCORRELATION
C       FUNCTION OF TEXTURE IMAGE BEFORE PUT IT IN MAX
C
C       INPUT IMAGE MUST STORE INTO ZIM.DAT
C
C       OCT.,9,1981
C       G.K.YOUNG
C
C++++++++++++++++++++++++++++++++++++++++++++++++++++
        INTEGER IF(64),N,N1,MN,MN2,M1,M2
        LOGICAL QGAP,GAP(25,25)
        REAL*4 F(64,64),TR(25,25),R(25,25)
        REAL*4 NOISE
        COMMON N,N2,N21,N4,N1,MN,MN2,MN3,MN4,M1,M2
        COMMON /LOGIC/GAP
        OPEN(UNIT=33,FILE='ZIM.DAT')
C
C       ASK FOR INPUT
C
        TYPE 12
12      FORMAT(5X,'DIMENSION OF ACF(MUST BE ODD)')
        ACCEPT 13,N12
13      FORMAT(I3)
        N1=(N12-1)/2
        MN2=13
        M1=MN2-N1
        M2=MN2+N1
        MN3=MN2+1
        MN4=2*MN2
        TYPE 14
14      FORMAT(5X,'DFT LENGTH(MUST BE EVEN)')
        ACCEPT 15,N
15      FORMAT(I3)
        N2=N/2
        N21=N2+1
        N4=N*N
        TYPE 16
16      FORMAT(5X,'NOISE POWER')
        ACCEPT 17,NOISE
17      FORMAT(F10.3)
        DO 1000 I=1,64
        DO 1005 L=1,64
1005    IF(L)=0
        DO 1010 J=1,4
        MI=J*16-15
        MJ=J*16
```

```
        READ(33,999)(IF(K),K=MI,MJ)
1010    CONTINUE
        DO 1020 M=1,64
        F(I,M)=FLOAT(IF(M))
1020    CONTINUE
1000    CONTINUE
999     FORMAT(16I4)
        CLOSE (UNIT=33,FILE='ZIM.DAT')
C
        OPEN(UNIT=31,FILE='ZTEM.DAT')
        DO 100 I=1,64
        DO 100  J=1,64
        F(I,J)=(F(I,J)-127.5)/127.5
100     CONTINUE
C
C       CALCULATE ACF
C
        DO 800 I=1,64
        NFAC=63*63
        MX=(N12+1)/2
        DO 500 I=1,N12
        DO 500 J=1,N12
        SUM=0.
        KI=IABS(I-MX)
        LJ=IABS(J-MX)
        DO 510 K=KI,63-KI
        DO 510 L=LJ,63-LJ
510     SUM=SUM+F(K,L)*F(K+I-MX,L+J-MX)
        TR(I,J)=SUM/NFAC
500     CONTINUE
        TRO=TR(MX,MX)
        DO 600 I=1,N12
        DO 600 J=1,N12
        TR(I,J)=TR(I,J)/TRO
600     CONTINUE
        DO 2222 I=1,N12
        WRITE(31,2000)(TR(I,J),J=1,N12)
111     FORMAT(5X,'R()= ',7F16.8,/,7F16.8,/,7F16.8)
2000    FORMAT(64F15.11)
2222    CONTINUE
        CLOSE (UNIT=31,FILE='ZTEM.DAT')
C
C       ADD NOISE
C
        IT=M1-1
        DO 40 I=M1,M2
        DO 40 J=M1,M2
        R(I,J)=TR(I-IT,J-IT)
40      CONTINUE
        DO 666 I=M1,M2
666     WRITE(5,111)(R(I,J),J=M1,M2)
        R(MN2,MN2)=R(MN2,MN2)+NOISE
        CALL EXIT
        END
```

```
C***********************************************************
C
C       MAX.FOR
C
C       PROGRAM TO OBTAIN THE 2-D MEM SOLUTION FOR REAL
C       DATA (REAL SYMMETRIC ACF), USING DIRECT DFT
C       COMPUTATION.
C
C       XLAM     =      LAMBDA COEFFICIENT ARRAY.
C       XOLD     =      OLD LAMBDA ARRAY REQUIRED FOR
C                       BETA COMPUTATION.
C       R        =      KNOWN ACF VALUE.
C       PX       =      COMPUTED ACF VALUES OR CORRECTION ACF.
C       GAP      =      LOGICAL ARRAY SPECIFYINGWHAT GAPS
C                       IF ANY, EXIST IN THE ACF
C       N        =      DFT LENGTH USED IN THE ITERATIONS.
C       N2       =      N/2
C       N21      =      N2+1
C       N4       =      N*N
C       N1       =      MAX +VE (X,Y) INDEX FOR THE SMALLEST
C                       SQUARE REGION CONTAINING 'R'.
C       MN       =      MAX ACF ARRAY SIZE FOR THIS PROGRAM(25).
C       MN2      =      CENTER POINT OF THE ACF ARRAY.
C       MN3      =      MN2+1
C       MN4      =      2*MN2
C       M1       =      MIN (XY) INDEX FOR KNOWN REGION.
C       M2       =      MAX (XY) INDEX FOR KNOWN REGION.
C       ALPHA    =      ALPHA IN THE ALGORITHM.
C       BETA     =      BETA  IN THE ALGORITHM.
C       SCLF     =      SCALE FACTOR, 'C' IN THE ALGORITHM.
C       ZTST     =      ERROR LEVAL.
C
C*************************************************************
        REAL*4 DEN
        REAL*4 X1,X2,Z
        REAL*4 XLAM(25,25),XOLD(25,25),R(25,25),PX(25,25)
        LOGICAL GAP(25,25)
C       DOUBLE PRECISION DPW
C
        COMMON  N,N2,N21,N4,N1,MN,MN2,MN3,MN4,M1,M2
        COMMON  /LOGIC/GAP
C       DEFINE FILE 1(628,4,U,NREC)
C
1       FORMAT(10X,' NITR = ',I3,5X,' ERROR = ',E12.4,//)
2       FORMAT(10X,' ALPHA = ',E14.6,//)
3       FORMAT(10X,' BETA = ',E14.6,//)
C
        MN=25
        MN2=(MN-1)/2+1
        WRITE(5,123)
123     FORMAT(15X,'*** RESULT OF 2-D MEM ***',///)
C
C       COMPUTE R AND SPECIFY GAPS, N,N1,M1,M2 ETC.
C       R(0,0)=1
```

```
      CALL ACF2D(R)
C
C     NORMALIZE R(MN2,MN2) TO 1
C
      RMN2=R(MN2,MN2)
C
      DO 35 I=M1,M2
      DO 35 J=M1,M2
      R(I,J)=R(I,J)/RMN2
35    CONTINUE
C
C     SPECIFY ERROR LEVAL, SCALE FACTOR AND INITIALIZE ARRAYS.
C
      ZOLD=1.0E30
      ZOLD1=1.0E30
      ZTST=1.0E-4
      SCLF=0.5
      ALPHA=0.0
      BETA=0.0
      DO 100 I=1,MN
      DO 100 J=1,MN
      XOLD(I,J)=0.0
      XLAM(I,J)=0.0
100   PX(I,J)=0.0
      XLAM(MN2,MN2)=1.0
      XOLD(MN2,MN2)=XLAM(MN2,MN2)
C
C     ERROR NORMALIZING FACTOR:
C
      DEN=0.0
      DO 110 I=M1,M2
      DO 110 J=M1,M2
      DEN=DEN+R(I,J)*R(I,J)
110   CONTINUE
      DEN=DEN-1.0
C
C     COMPUTE CORRESPONDING ACF VALUES FOR REGION 'A' ONLY.
C     ALSO OBTAINED ARE THE MIN(F(\A]BDA)) AND MIN(F(R)),
C     R'(0,0)=1.0
      NITR=0
1000  CALL FT(PX,XLAM,X1,X2)
      NITR=NITR+1
C
C     OBTAIN CORRECTION ACF AND ACF ERROR
C
      Z=0.0
      DO 120 I=M1,M2
      DO 120 J=M1,M2
      PX(I,J)=R(I,J)-PX(I,J)
120   Z=Z+PX(I,J)*PX(I,J)
C
C     NORMALIZE ACF ERROR.
C
      Z=Z/DEN
```

```
      WRITE(5,1)NITR,Z
      IF (Z.LT.ZTST) GO TO 9000
      IF (ALPHA.GT.0.9999) GOTO 9000
      IF (Z.GE.ZOLD.OR.BETA.NE.0.0) GO TO 970
      GO TO 971
970   ALPHA=(1.0+ALPHA)/2.0
      SCLF=SCLF/2.0
      ZOLD1=ZOLD
971   ZOLD=Z
C     IF(Z.GE.ZOLD.AND.ZOLD.GE.ZOLD1)GO TO 9999
C
C     COMPUTE MIN VALUE OF THE CORRECTION SPECTRUM
C     AND UPDATE ALPHA, IF NECESSARY.
C
      CALL FT2(PX,X3)
      XX=1.0+SCLF*X2/X3
      ALPHA=AMAX1(ALPHA,XX)
      WRITE(5,2)ALPHA
C
C     COMPUTE LAMBDA VALUES FOR REGION 'A' ONLY
C
      CALL FT3(XOLD,PX,ALPHA,XLAM)
C
C     COMPUTE MIN VALUE OF F(LAMBDA)
C
      CALL FT2(XLAM,X4)
      IF (X4.GT.0.0) GO TO 972
      BMIN=-X4/(X1-X4)
      BETA=(1.0+(1.0-SCLF)*(1.0/BMIN-1.0))*BMIN
      ALPHA=(1.0+ALPHA)/2.0
      GO TO 973
972   BETA=0.0
973   CONTINUE
      WRITE(5,3)BETA
C
C     OBTAIN NEW LAMBDA SET
C
      DO 140 I=M1,M2
      DO 140 J=M1,M2
      XLAM(I,J)=BETA*XOLD(I,J)+(1.0-BETA)*XLAM(I,J)
140   XOLD(I,J)=XLAM(I,J)
      GO TO 1000
9000  CONTINUE
C
C     CONVERGENCE ACHIEVED, TEST FOR ERROR ON A FINER GRID
C
      TYPE 9900
9900  FORMAT(1X,' CONVERGENCE ACHIEVED! TEST IN PROGRESS,
     * PATIENCE.... ')
      N=4*N
      N2=N/2
      N21=N2+1
      N4=N*N
      CALL FT(PX,XLAM,X1,X2)
```

```
C       IF F(LAMBDA)<0, ERROR MESSAGE ISSUED BY FT.  IF
C       >0 THEN NEW ERROR LEVAL IS CHECKED BELOW.
C
        Z=0.0
        DO 9010 I=M1,M2
        DO 9010 J=M1,M2
9010    Z=Z+(R(I,J)-FX(I,J))**2
        Z=Z/DEN
9001    FORMAT(' GOOD SOLUTION. ERROR IS: ',E12.4,///)
9002    FORMAT(' ERROR LEVAL TOO HIGH: ',E12.4)
        IF (Z.LE.ZTST) GO TO 974
        WRITE(5,9002)Z
        GO TO 975
974     WRITE(5,9001)Z
975     CONTINUE
C
C       WRITE OUT FILTER INTO FILE.
C
        OPEN(UNIT=30,FILE='ZFIL.DAT')
        LACF=M2-M1+1
C       DPW=DBLE(FLOAT(LACF))
        DPW=LACF
        NREC=1
        WRITE(30,19)DPW
19      FORMAT(F18.10)
C       WRITE(1'NREC)DPW
        DPW=1.0
        NREC=2
        WRITE(30,19)DPW
C       WRITE(1'NREC)DPW
        NREC=3
        DO 9020 I=M1,M2
        DO 9020 J=M1,M2
        WRITE(5,9003)I,J,XLAM(I,J)
9003    FORMAT(1X,' LAMBDA(',I2,1X,I2,')= ',E14.6)
C       DPW=DBLE(XLAM(I,J))
        DPW=XLAM(I,J)
        WRITE(30,19)DPW
C       WRITE(1'NREC)DPW
9020    CONTINUE
        DPW=1.0
C       WRITE(1'NREC)DPW
        WRITE(30,19)DPW
        CLOSE(UNIT=30,FILE='ZFIL.DAT')
C       WRITE OUT ACF(R') ARRAY INTO FILE 2
C
        STOP
C       CALL EXIT
        END
C***********************************************************
        SUBROUTINE ACF2D(R)
C***************************************************************
C       LOGICAL QGAP
        LOGICAL GAP(25,25)
```

```
      INTEGER N,N1,MN,MN2,M1,M2
      REAL*4 R(25,25),P(20),XFREQ(20),YFREQ(20),NOISE,T(25,25)
      COMMON N,N2,N21,N4,N1,MN,MN2,MN3,MN4,M1,M2
      COMMON /LOGIC/GAP
      PI = 4.0 * ATAN(1.0)
C     DEFINE FILE 3(25,50,U,LREC)
C
C   ASK FOR INPUT
C
      TYPE 110
110   FORMAT(10X,' NUMBER OF SINUSOIDS ')
      ACCEPT 111,NSIN
111   FORMAT(I2)
      WRITE(5,112)NSIN
112   FORMAT(5X,' NUMBER OF SINUSOIDS = ',I2)
C
      DO 10 I=1,NSIN
      TYPE 130
130   FORMAT(10X,' POWER,XFREQ,YFREQ( ABS(FREQ) =< 0.5 ')
      ACCEPT *,P(I),XFREQ(I),YFREQ(I)
131   FORMAT(F8.2,1X,F7.5,1X,F7.5)
      WRITE(5,132)I,P(I),XFREQ(I),YFREQ(I)
132   FORMAT(5X,' SINUSOID ',I3,5X,' POWER = ',F8.2,5X,
     * ' XFREQ = ',F7.5,5X,' YFREQ = ',F7.5)
10    CONTINUE
C
      TYPE 150
150   FORMAT(10X,' NOISE POWER')
      ACCEPT 151,NOISE
151   FORMAT(F7.3)
      WRITE(5,152)NOISE
152   FORMAT(5X,' NOISE POWER = ',F7.3)
C
      TYPE 170
170   FORMAT(10X,' DIMENSION OF SMALLEST SQUARE CONTAINING
     * ACF (MUST BE ODD)')
      ACCEPT 171,N12
171   FORMAT(I2)
      WRITE(5,172)N12,N12
172   FORMAT(5X,' ACF MATRIX ',I2,'X',I2)
C
      N1 = (N12-1)/2
      M1 = MN2 - N1
      M2 = MN2 + N1
      MN3= MN2 + 1
      MN4= 2*MN2
C   FILL GAP ARRAY
C
      DO 15 I=M1,M2
      DO 15 J=M1,M2
15    GAP(I,J)=.TRUE.
C
      TYPE 180
180   FORMAT(10X,' ANY GAPS? ')
```

```
      ACCEPT 181,QGAP
181   FORMAT(F3.2)
      IF (QGAP.NE.1.) GO TO 980
      TYPE 190
190   FORMAT(10X,'NUMBER OF GAPS ')
      ACCEPT 191,NGAP
191   FORMAT(I2)
      DO 25 K=1,NGAP
      TYPE 193
193   FORMAT(10X,' MATRIX COORDINATES: I,J ')
      ACCEPT 194,I,J
194   FORMAT(2I3)
      WRITE(5,195)I,J
195   FORMAT(5X,' R(',I3,',',I3,') IS A GAP')
25    GAP(M1+I-1,M1+J-1) = .FALSE.
980   CONTINUE
      TYPE 197
197   FORMAT(10X,' DFT LENGTH (MUST BE EVEN)')
      ACCEPT 198,N
198   FORMAT(I5)
      WRITE(5,199)N
199   FORMAT(5X,' DFT LENGTH = ',I5,////)
      N2 = N/2
      N21 = N2+1
      N4 = N*N
C   CALCULATE ACF
C
      LREC=1
C     DO 30 L=1,N12
C     READ(3'LREC)(T(L,M),M=1,N12)
      DO 30 NS=1,NSIN
             WX=2.*PI*XFREQ(NS)
             WY=2.*PI*YFREQ(NS)
        DO 30 I=M1,M2
             IA=I-MN2
        DO 30 J=M1,M2
             JA=J-MN2
        IF(NS.EQ.1) R(I,J)=0.0
30    IF(GAP(I,J).EQ..TRUE.) R(I,J)=R(I,J)+P(NS)*COS(WX*IA+WY*JA)
C     IT=M1-1
      DO 40 I=M1,M2
C     DO 40 J=M1,M2
C     R(I,J)=T(I-IT,J-IT)
40    CONTINUE
C
C   ADD NOISE
C
      DO 50 I=M1,M2
C     WRITE(5,44)(R(I,J),J=M1,M2)
50    CONTINUE
44    FORMAT(1X,7F16.8,/,7F16.8,/,7F16.8)
      R(MN2,MN2) = R(MN2,MN2) + NOISE
      RETURN
      END
```

```
      SUBROUTINE FT2(PX,XMIN)
C*************************************************************C
C     SUBROUTINE TO COMPUTE THE MINIMUM VALUE OF
C     THE FOURIER TRANSFORM OF PX.
C
C*************************************************************
      REAL*4 PX(25,25)
      COMPLEX Z0,Z1,Z2,Z3,Z4,Z5,Z6,Z7
      LOGICAL GAP(25,25)
C
      COMMON N,N2,N21,N4,N1,MN,MN2.MN3,MN4,M1,M2
      COMMON /LOGIC/GAP
C
      PIN=8.0*ATAN(1.0)/FLOAT(N)
C
C   COMPUTE THE (0,0) AND (N/2,N/2) VALUES.
C
      X00=0.0
      X22=0.0
      DO 100 I=M1,M2
      DO 100 J=M1,M2
      X00=X00+PX(I,J)
100   X22=X22+PX(I,J)*((-1.0)**(I+J-MN4))
      XMIN=AMIN1(X00,X22)
C
C
C     COMPUTE ALL OTHER INDEPENDENT X(K,L) VALUES
C     IN THE UPPER HALF PLANE, USING RECURSIVE
C     COSINE COMPUTATION.
C
C
      X0=PX(MN2,MN2)
      Z0=CMPLX(COS(PIN),SIN(PIN))
      Z1=(1.0,0.0)
      Z6=CMPLX(COS(PIN*N1),-SIN(PIN*N1))
      Z7=(1.0,0.0)
C
C     SPECIFY (K,L) VALUES IN THE UPPER HALF PLANE.
C
      DO 600 K=1,N21
C
C     CONTRIBUTION TO XKL OF THE 0'TH PX COLUMN.
C
      Z2=Z1
      X1=0.0
      DO 200 M=MN3,M2
      X1=X1+PX(M,MN2)*REAL(Z2)
200   Z2=Z2*Z1
C
      Z3=(1.0,0.0)
      DO 500 L=1,N
      IF(K*L.EQ.1) GO TO 500
      IF(K.EQ.1.AND.L.GT.N21) GO TO 601
      IF(K.EQ.N21.AND.L.GT.N2) GO TO 600
```

```
      X2=0.0
      Z4=Z7
      DO 400 M=M1,M2
      Z5=Z3
      DO 300 NN=MN3,M2
      X2=X2+PX(M,NN)*REAL(Z4*Z5)
300   Z5=Z5*Z3
      IF(M.EQ.M2) GO TO 400
      Z4=Z4*Z1
400   CONTINUE
C
      XKL=X0+2.0*(X1+X2)
      XMIN=AMIN1(XMIN,XKL)
500   Z3=Z3*Z0
601   Z1=Z1*Z0
      Z7=Z6*CONJG(Z4)
600   CONTINUE
      RETURN
      END
C***********************************************************C
      SUBROUTINE FT3(XOLD,PX,ALPHA,XLAM)
C************************************************************
C
C     SUBROUTINE TO COMPUTE THE NEW LAMBDA VALUES FROM THE
C     CORRECTED ACF. ONLY HALF THE ARRAY IS NEEDED DUE TO
C     CORRECTED ACF. ONLY HALF THE ARRAY IS NEEDED DUE TO
C     THE SYMMETRY.
C     RECURSIVE COSINE COMPUTATION.
C
C************************************************************
C
      REAL*4 PX(25,25),XLAM(25,25),XOLD(25,25)
      LOGICAL GAP(25,25)
      COMPLEX Z0,Z1,Z2,Z3,Z4,Z5,Z6,Z7
C
      COMMON N,N2,N21,N4,N1,MN,MN2,MN3.MN4,M1,M2
      COMMON /LOGIC/GAP
C
      XN2=N4
      PIN=8.0*ATAN(1.0)/FLOAT(N)
C
C     COMPUTE THE (0,0) AND (N/2,N/2) VALUES
C
      X00=0.0
      X22=0.0
      Y00=0.0
      Y22=0.0
      DO 100 I=M1,M2
      DO 100 J=M1,M2
      X00=X00+PX(I,J)
      X22=X22+PX(I,J)*((-1.0)**(I+J-MN4))
      Y00=Y00+XOLD(I,J)
      Y22=Y22+XOLD(I,J)*((-1.0)**(I+J-MN4))
100   XLAM(I,J)=0.0
```

```
      X00=1.0/Y00+(1.0-ALPHA)*X00
      X00=1.0/X00
      X22=1.0/Y22+(1.0-ALPHA)*X22
      X22=1.0/X22
C
C     PLACE THE IDFT CONTRIBUTION OF X00 & X22 IN XLAM
C
      DO 150 M=M1,M2
      IF(M.LT.MN2) GO TO 985
      NN1=MN2
      GO TO 986
985   NN1=MN3
986   CONTINUE
      DO 150 NN=NN1,M2
150   XLAM(M,NN)=X00+X22*((-1.0)**(M+NN-MN4))
C
C     COMPUTE ALL OTHER INDEPENDENT X(K,L) AND Y(K,L)
C     VALUES IN THE UPPER HALF PLANE. USING RECURSIVE
C     COSINE COMPUTATION.
C
      X0=PX(MN2,MN2)
      Y0=XOLD(MN2,MN2)
      Z0=CMPLX(COS(PIN),SIN(PIN))
      Z1=(1.0,0.0)
      Z6=CMPLX(COS(PIN*N1),-SIN(PIN*N1))
      Z7=(1.0,0.0)
C
C     SPECIFY (K,L) VALUES IN THE UPPER HALF PLANE:
C
      DO 800 K=1,N21
C
C     CONTRIBUTION TO XKL OF THE 0'TH PX AND XOLD COLUMNS.
C
      Z2=Z1
      X1=0.0
      Y1=0.0
      DO 200 M=MN3,M2
      XX=REAL(Z2)
      X1=X1+PX(M,MN2)*XX
      Y1=Y1+XOLD(M,MN2)*XX
200   Z2=Z2*Z1
C
      Z3=(1.0,0.0)
      DO 700 L=1,N
      IF(K*L.EQ.1) GO TO 700
      IF(K.EQ.1.AND.L.GT.N21) GO TO 801
      IF(K.EQ.N21.AND.L.GT.N2) GO TO 800
C
      X2=0.0
      Y2=0.0
      Z4=Z7
      DO 400 M=M1,M2
      Z5=Z3
```

```
      DO 300 NN=MN3,M2
      XX=REAL(Z4*Z5)
      X2=X2+PX(M,NN)*XX
      Y2=Y2+XOLD(M,NN)*XX
300   Z5=Z5*Z3
      IF(M.EQ.M2) GO TO 400
      Z4=Z4*Z1
400   CONTINUE
C
      XKL=X0+2.0*(X1+X2)
      YKL=Y0+2.0*(Y1+Y2)
      XKL=1.0/YKL+(1.0-ALPHA)*XKL
      XKL=1.0/XKL
C
C     GIVEN THE XKL VALUE,UPDATE THE RUNNING SUMS FOR
C     THE NEW LAMBDA COEFFICIENTS:
C
      Z2=Z7
      DO 600 M=M1,M2
      IF(M.LT.MN2) GO TO 987
      Z5=(1.0,0.0)
      NN1=MN2
      GO TO 988
987   Z5=Z3
      NN1=MN3
988   CONTINUE
      DO 500 NN=NN1,M2
      IF(.NOT.GAP(M,NN)) GO TO 500
      IF(K*L.EQ.N21) GO TO 989
      XLAM(M,NN)=XLAM(M,NN)+2.0*XKL*REAL(Z2*Z5)
      GO TO 500
989   XLAM(M,NN)=XLAM(M,NN)+XKL*REAL(Z2*Z5)
500   Z5=Z5*Z3
600   Z2=Z2*Z1
700   Z3=Z3*Z0
801   Z1=Z1*Z0
      Z7=Z6*CONJG(Z4)
800   CONTINUE
C
C     HALF ARRAY OF LAMBDA VALUES COMPUTED. REARRANGE TO FULL
C     ARRAY AND NORMALIZE.
C
      DO 900 M=M1,M2
      DO 900 NN=M1,MN2
900   XLAM(M,NN)=XLAM(M1+M2-M,M1+M2-NN)
C
      DO 1000 M=M1,M2
      DO 1000 NN=M1,M2
1000  XLAM(M,NN)=XLAM(M,NN)/XN2
      RETURN
      END
```

```
C*************************************************************C
        SUBROUTINE FT(PX,XLAM,XMIN1,XMIN2)
C*************************************************************C
C       SUBROUTINE TO COMPUTE THE ACF VALUES FROM THE LAMBDA
C       COEFFICIENTS VIA DIRECT DFT COMPUTATION. SYMMETRY IS
C       FULLY USED. REAL, SYMMETRIC ACF AND LAMBDA AND HENCE
C       F(LAMBDA)=1/F(ACF).
C       RECURSIVE COSINE COMPUTATION.
C*************************************************************C
        REAL*4 PX(25,25),XLAM(25,25),XMIN1,XMIN2
        LOGICAL GAP(25,25)
        COMPLEX Z0,Z1,Z2,Z3,Z4,Z5,Z6,Z7
C
        COMMON N,N2,N21,N4,N1,MN,MN2,MN3,MN4,M1,M2
        COMMON /LOGIC/GAP
C
        XN2=N4
        PIN=8.0*ATAN(1.0)/FLOAT(N)
C
C       COMPUTE THE (0,0) AND (N/2,N/2) VALUES.
C
        X00=0.0
        X22=0.0
        DO 100 I=M1,M2
        DO 100 J=M1,M2
        X00=X00+XLAM(I,J)
        X22=X22+XLAM(I,J)*((-1.0)**(I+J-MN4))
100     PX(I,J)=0.0
        IF(X00.LT.0.0.OR.X22.LE.0.0) GO TO 990
        GO TO 991
990     TYPE 11
11      FORMAT(5X,'BAD SOLUTION, F(LAMBDA) < 0')
        STOP
991     CONTINUE
        XMIN1=AMIN1(X00,X22)
        XMIN2=AMAX1(X00,X22)
        X00=1.0/X00
        X22=1.0/X22
C
C       PLACE THE IDFT CONTRIBUTION ON X00 & X22 IN PX
C
        DO 150 M=M1,M2
        IF(M.LT.MN2) GO TO 992
        NN1=MN2
        GO TO 993
992     NN1=MN3
993     CONTINUE
        DO 150 NN=NN1,M2
150     PX(M,NN)=X00+X22*((-1.0)**(M+NN-MN4))
C
C       COMPUTE ALL OTHER INDEPENDENT X(K,L) VALUES
C       IN THE UPPER HALF PLANE. USING RECURSIVE
C       COSINE COMPUTATION.
```

```
C
      X0=XLAM(MN2,MN2)
      Z0=CMPLX(COS(PIN),SIN(PIN))
      Z1=(1.0,0.0)
      Z6=CMPLX(COS(PIN*N1),-SIN(PIN*N1))
      Z7=(1.0,0.0)
C
C     SPECIFY (K,L) VALUES IN THE UPPER HALF PLANE:
C
      DO 800 K=1,N21
C
C     CONTRIBUTION TO XKL OF THE 0'TH XLAM COLUMN
C
      Z2=Z1
      X1=0.0
      DO 200 M=MN3,M2
      X1=X1+XLAM(M,MN2)*REAL(Z2)
200   Z2=Z2*Z1
C
      Z3=(1.0,0.0)
      DO 700 L=1,N
      IF(K*L.EQ.1) GO TO 700
      IF(K.EQ.1.AND.L.GT.N21) GO TO 801
      IF(K.EQ.N21.AND.L.GT.N2) GO TO 800
C
      X2=0.0
      Z4=Z7
      DO 400 M=M1,M2
      Z5=Z3
      DO 300 NN=MN3,M2
      X2=X2+XLAM(M,NN)*REAL(Z4*Z5)
300   Z5=Z5*Z3
      IF (M.EQ.M2) GO TO 400
      Z4=Z4*Z1
400   CONTINUE
C
      XKL=X0+2.0*(X1+X2)
      IF(XKL.GT.0.0) GO TO 994
      TYPE 222
222   FORMAT(5X,' BAD SOLUTION, F(LAMBDA) < 0')
      STOP
994   CONTINUE
      XMIN1=AMIN1(XMIN1,XKL)
      XMIN2=AMAX1(XMIN2,XKL)
      XKL=1.0/XKL
C
C     GIVEN THE XKL VALUE, UPDATE THE RUNNING SUMS
C     FOR THE ACF R'(M,N)
C
      Z2=Z7
```

```
      DO 600 M=M1,M2
      IF(M.LT.MN2) GO TO 995
      Z5=(1.0,0.0)
      NN1=MN2
      GO TO 996
995   Z5=Z3
      NN1=MN3
996   CONTINUE
      DO 500 NN=NN1,M2
      IF(.NOT.GAP(M,NN)) GO TO 500
      IF(K*L.EQ.N21) GO TO 997
      PX(M,NN)=PX(M,NN)+2.0*XKL*REAL(Z2*Z5)
      GO TO 500
997   PX(M,NN)=PX(M,NN)+XKL*REAL(Z2*Z5)
500   Z5=Z5*Z3
600   Z2=Z2*Z1
700   Z3=Z3*Z0
801   Z1=Z1*Z0
      Z7=Z6*CONJG(Z4)
800   CONTINUE
C
C     HALF ARRAY OF ACF VALUES COMPUTED. REARRANGE TO FULL
C     ARRAY AND NORMALIZE.
C
      DO 900 M=M1,M2
      DO 900 NN=M1,MN2
900   PX(M,NN)=PX(M1+M2-M,M1+M2-NN)
C
      PX0=PX(MN2,MN2)
      XX=PX0/XN2
      DO 1000 M=M1,M2
      DO 1000 NN=M1,M2
      PX(M,NN)=PX(M,NN)/PX0
1000  XLAM(M,NN)=XLAM(M,NN)*XX
C
      XMIN2=1.0/(XMIN2*XX)
      XMIN1=XMIN1*XX
      RETURN
      END
```

Index